Congratulations

Charlie D Reen

Material Handling Systems

Designing for Safety and Health

Material Handling Systems

Designing for Safety and Health

Charles D. Reese

Taylor & Francis
New York

USA	Publishing Office	Taylor & Francis
		29 West 35th Street
		New York, NY 10001-2299
		Tel: (212)216-7800
	Distribution Center	Taylor & Francis
		47 Runway Road, Suite G
		Levittown, PA 19057-4700
		Tel: (215)269-0400
		Fax: (215)269-0363
UK		Taylor & Francis
		11 New Fetter Lane
		London EC4P 4EE
		Tel: 011 44 207 583 9855
		Fax: 011 44 207 842 2298

MATERIAL HANDLING SYSTEMS: DESIGNING FOR SAFETY AND HEALTH

1 2 3 4 5 6 7 8 9 0

A CIP catalog record for this book is available from the British Library.

Library of Congress Cataloging-in-Publication Data

Reese, Charles D.
   Material handling systems : designing for safety and health / by Charles Reese.
     p. cm
   ISBN 1-56032-868-1 (alk. paper)
   1. Materials handling—Safety measures. I. Title.

TS180 .R435 2000
621.8'6—dc2l

                                                 99-057159

DEDICATION

This book is dedicated to all those who saw in me potential and gave me the opportunity to accomplish so much. To Carol, my wife who shakes her head but stands behind me when I try something new. To Craig and Chad, my sons who have always been interested and supportive of my travels through this life.

TABLE OF CONTENTS

CHAPTER 9 RIGGING 125

ABOUT THE AUTHOR

For over twenty years Dr. Charles D. Reese has been involved with occupational safety and health as an educator, manager, or consultant. In Dr. Reese's early beginnings in occupational safety and health, he held the position of industrial hygienist at the National Mine Health and Safety Academy. He later assumed the responsibility of manager for the nation's occupational trauma research initiative at the National Institute for Occupational Safety and Health's (NIOSH) Division of Safety Research. Dr. Reese has had an integral part in trying to assure that workplace safety and health is provided for all those within the workplace. As the managing director for the Laborers' Health and Safety Fund of North America, his responsibilities were aimed at protecting the 650,000 members of the laborers' union in the United States and Canada.

He has developed many occupational safety and health training programs which run the gamut from radioactive waste remediation to confined space entry. Dr. Reese has written numerous articles, pamphlets, and books on related safety and health issues.

At present Dr. Reese is a member of the graduate and undergraduate faculty at the University of Connecticut, where he teaches courses on OSHA regulations, safety and health management, accident prevention techniques, industrial hygiene, and ergonomics. As associate professor of occupational safety and health, he coordinates the bulk of the safety and health efforts at the University and Labor Education Center. He is often called upon to consult with industry on safety and health issues and also asked for expert consultation in legal cases.

Dr. Reese is also the principal author of the *Handbook of OSHA Construction Safety and Health.*

PREFACE

Material Handling Systems: Designing for Safety and Health was developed and written as a comprehensive approach to occupational safety and health related to the myriad of material handling approaches which are in use by a wide variety of industrial sectors.

The emphasis is primarily upon safety, and the prevention of those accidents, injuries, and deaths which transpire with a high degree of frequency. More precisely, those events which occur in the process of material handling are addressed.

This book is an attempt to point out the safety and health concerns as well as the regulatory requirements for safe material handling. Many material handling venues have been discussed, from cranes to industrial robots. Certainly, this diverse approach to material handling safety will be of interest to many who are responsible for safety or having material handling as a major component of their operation. Suffice it to say that material handling activities impact most workplaces in some manner.

ACKNOWLEDGEMENTS

I would like to give special thanks to John Palumbo, my graduate assistant, who gave all of the chapters a first read to catch my many errors.

Again, I would be remiss to overlook another great work and effort by Kay Warren of BarDan Associates, who put this manuscript into camera ready copy.

I would like to recognize the editing work of Anita B. Kersten of Beaupré & Associates, who diligently edited my initial draft of this manuscript.

Also, I appreciate the courtesy extended to me by the following organizations and individuals:

American Society of Mechanical Engineers
The Crosby Group, Inc.
Kaman Corporation
National Institute for Occupational Safety and Health
Occupational Health and Safety Administration
Rite-Hite Corporation
The Aldon Co.
United Abrasive, Inc.
United States Department of Energy

Lawrence L. Jaglowski
Charles A. O'Neal
Joseph Petrello
John Tabak
Charles J. Woods

CHAPTER 1

INTRODUCTION TO MATERIAL HANDLING

Material handling of sand by heavy equipment

    Material Handling Systems: Designing for Safety and Health is envisioned to be both a practical guide and instructional tool with application to all industries, which are faced with the handling of materials, and to those individuals who have responsibility for occupational safety and health. This book covers the many facets of safety and health related to hazards faced by work areas and workers who must lift, move, or in some way handle a variety of materials.

    This book provides a detailed approach to the topic of material handling, which impacts most workplaces in one way or another, whether it be the handling of material on a production line, the movement of heavy equipment, or the warehousing and storage of supplies or products.

    The handling of all types of materials may manifest itself in the individual worker's effort to lift or move materials using large industrial cranes. No matter which procedure is used, there are hazards and safety concerns, which need to be addressed. Almost every industrial sector has to address material handling issues, especially workplaces moving materials in and products out on a just-in-time schedule. Yet improper handling and storage of materials can result in costly injuries. Materials may be anything from boxes, parts, or equipment to steel beams, aircraft engines, or manufactured homes.

The efficient handling and storage of materials is vital to the function of industry. Material handling operations provide for the continuous flow of raw materials, parts, and products throughout the workplace and assure that materials and products are there when they are needed. Yet the improper handling and storage of materials can cause costly injuries.

The proper and safe handling of a wide variety of materials must be done in compliance with existing Occupational Safety and Health Administration (OSHA) regulatory guidelines for the equipment used, the methods or procedures followed, and the appropriate storage of each type of material.

This book incorporates all facets of material handling as it applies to the workplace. It addresses the real issues of safe handling of materials faced by employers and their employees. More employees are injured in industry while moving materials than performing any other single function. These injuries have been estimated to account for 20 percent to 25 percent of all occupational injuries. Readers are provided the tools for addressing material handling problems, information on regulatory requirements, and instruction on safe use of material handling equipment in an effort to prevent and decrease these types of injuries.

The text discusses the most common safety and health hazards present during material handling activities including controls, tools for prevention, safety equipment use including cranes and forklifts, and how to comply with the regulatory requirements. Help will be provided in the form of checklists, informational sources, and resources available to assist in addressing concerns related to material handling through safety and health. This book has been written primarily for industry employers, managers, workers, safety and health professionals, and academics who have an interest in safety and health. It a good source for industrial and construction engineering students and other engineering students who may need to address material handling issues.

Improper material handling is one of the most frequently results in hazards ranging from back injury to events as catastrophic as the collapse of a crane, which could result in loss of lives and extensive property damage. When an average back injury costs $9,000 (the most common body part injured by workers) and the cost of one fatality is approximately $1,000,000, most companies cannot absorb these types of financial losses or the loss of resources. These resources are in the form of lost production, human suffering, and real dollar costs which quickly get the attention of owners and investors as well as workers.

These soaring costs have triggered attention to occupational safety and health beyond the regulatory requirements of OSHA. It seems safe to say that material handling safety and health in the workplace are everybody's business and can have a detrimental effect on the bottom line. Thus, any attention paid to safe handling of materials will result in a payoff. Using this text, readers will be able to adapt and implement material handling safety policies and health initiatives in their workplace, department, or job.

This book is to be used as a source for safe handling of materials for those who are attempting to improve the overall safety and health performance at an existing company as well as a resource for those individuals who have been assigned responsibility for safety and health for a company. Also, it has application to academics from safety and healthy programs and to engineering students. In engineering curricula, safety and health has often been neglected. Since this text applies to all industry-specific safety and health material handling, it should match better with the engineering curriculum where handling of materials, parts, products, and equipment are at issue.

This book is to be a practical guide with application to all industries and to companies of all sizes in the critical area of occupational safety and health related to the handling of materials. (See Figure 1-1.)

Figure 1-1. Material handling at its basic level

When one thinks of the ominous task of trying to effectively address the needs of those faced with material handling issues, it may seem almost insurmountable. Estimates by OSHA and the Bureau of Labor Statistics (BLS) suggest that in the United States there are at least 6.5 million different workplaces and some 100 million workers who suffer 6 million injuries per year. Because there are so many different workplaces, this makes for a great deal of diversity when one considers material handling issues.

Handling and storage of material involve diverse operations such as hoisting tons of steel with a crane, stacking products in a warehouse with a forklift, off loading a highway truck, manually carrying bags and materials, stacking barrels, drums, or kegs, delivering a load of concrete blocks, or the stacking and storage of loose materials, lumber, or pipes.

HAZARDS INVOLVED

Injuries faced by those performing material handling tasks may be something as simple as overexertion, which results in sprains or strains, or simple cuts and lacerations from sharp edges or contact with moving parts on equipment. The pinch (nip) points or sheer weight of items being handled can result in bruises, contusions, crushing, fractures, and amputations. The larger the object, the larger the equipment being used, and the faster the movement of materials, the greater the risk for multiple injuries, suffocation, or, worst of all, death. (See Figure 1- 2.)

Many of the materials being handled by workers include hazardous chemicals, which may present both an injury and illness potential. Chemicals can cause fires or explosion hazards and can result in burns or concussion injuries. Others may present the potential for contact, ingestion, or inhalation exposures, which may cause allergic reactions or toxic (poisonous) effects in workers, when such materials are mishandled and/or spilled.

It certainly seems safe to say that all of these scenarios have transpired at one time or another to workers handling hazardous chemicals. It appears worthwhile to expand this discussion of material handling hazards to the accidents/incidents caused by them. Certainly, if

Figure 1- 2. The risk of injury can be high æ while handling awkward materials

some materials are too heavy and when a lot of repetitious lifting occurs, the potential for overexertion will likely ensue with sprains and strains.

Materials which are improperly stored or handled have the potential to shift due to their weight, shape, or potential to flow. For example, sand being moved and stored is not at its normal angle of repose and may engulf a worker. This is particularly a problem around stockpiles, surge bins, or excavations. When material shifts, it may physically strike a worker, pinning him/her between a stationary object and the moving materials.

When using the wide variety of equipment available to move or handle the different types of materials that exist in the workplace, the unevenness, unsecured loads, and extreme weight of the loads being lifted or moved can cause equipment to malfunction, collapse or, at least, function erratically. The load can potentially fall, striking a worker, swing into a worker, be caught under a piece of equipment, or under a load, which cannot be controlled. This is why it is important that proper inspections and maintenance must be performed on equipment used for handling material. If a sling (wire rope, steel alloy chain, or webbed sling) fails, a crane boom collapses, or the brakes fail on a vehicle, the end results can be disastrous.

The use of equipment to handle materials is controlled by pre-established lifting or load limits and restrictions on the supporting capacity of storage (shelving) units which can never be exceeded if safety is a primary focus. Assurance must exist that the appropriate equipment is used for the job, and that it is properly used by the operator. If an operator inadvertently contacts an electrical conductor, for example, electrocution is a real possibility.

Workers do not expect to be working under a load, have a load fall on them from above, or to be run over by a piece of material handling equipment. These hazards are preventable using fundamental safety precautions.

The last hazard is derived from the myriad of chemicals handled or stored within the workplace. Not only do they present the potential to cause physical harm as noted earlier when chemical containers shift, roll, or strike a worker, injuring or killing that worker, many chemicals pose another type of hazard. They can potentially be toxic (poisonous) or cause burns if mishandled or spilled and are not properly controlled. Some chemicals may also create an explosion or fire.

As can be seen, the movement, stacking, and storage of materials pose many hazards within the workplace. The philosophical approach to these hazards must be that we can identify them, they are preventable, and we can significantly reduce the accidents/incidents, which result in injuries, illnesses, and deaths from improper material handling.

TYPES OF EQUIPMENT BEING USED TODAY

Probably few facets of safety and health have such a wide variety of equipment with which to be concerned as material handling. Some equipment is as simple as a handcart or dolly, or complex as industrial robots. Examples of equipment which is usually categorized as non-powered are shown in Table 1-1 and Figure 1-3.

A mechanical advantage can be enhanced by using powered equipment for smaller items not requiring industrial hoists, forklifts, cranes, or heavy-duty trucks. Some of these types of handling devices which are available are operated by hydraulic, compressed air, or electrical energy. Examples of powered equipment are found in Table 1-2 and Figure 1-4.

Table 1-1

Non-powered Equipment

Block and tackle	Service,
Carts,	bench
basket	Slings,
instrument	alloy steel chain
service	webbed
stock	wire rope
utility	Stairs,
Chain hoist	rolling
Conveyors,	Straps
roller	Trucks,
skate wheel	adjustable tray
Derrick,	bar
roof	deck
Dockboards	drum handling
Dollies	hand
Gantry crane	package
Handles	platform
Hose reels	shelf
Ladders,	sheet
rolling	tilt
	tube
Ramps	wagon
	Wire dispenser

Figure 1-3. Use of a hand truck and ramp for delivery

Table 1-2

Powered Material Handling Equipment

Air or electric hoist	Lift tables (scissors lift tables)
Drum stackers	Material lifting trucks (lift trucks)
Dumping devices for hoppers and drums	Pallet truck
Floor crane	Power gantry crane (light weight)
High reach work platforms	Power operated conveyor (belt)
Jacks	Stock pickers
Jib cranes	Tilt tables
Powered loading docks	Vacuum Lift
	Work positioners

Many other forms of material handling equipment are used with extremely heavy loads. Most are very sophisticated and require both that operators be highly trained and that workers around such large and powerful equipment know the inherent hazards.

At times, the size of equipment used to handle heavy loads varies greatly. A situation may call for a standard dump truck or a mega truck such as those used on mining and construction sites. Examples of equipment used for transport and haulage are found in Table 1-3 and Figure 1-5.

Figure 1-4. Use of a scissor lift for material handling

Table 1-3

Vehicles Used for Transport and Haulage

The industrial trucks (forklifts)
Highway/off-highway dump trucks
Long-haul trucks and trailers
Industrial cranes, derricks and hoists
Excavators, dozers, scrapers, front-end loader,
backhoes, shovels, and dredges

The most sophisticated of material handling equipment is the industrial robot, which is designed to replace the worker when high-speed, repetitive, and consistent performance and accuracy are desirable. The close tie of robots to electronic and computer controlled operations presents a unique set of problems when used in the workplace for material handling activities.

As noted, material handling equipment constitutes a continuum from the most primitive lever devices (still quite useful) to the state-of-the art modern industrial robots. Again, each type of equipment, from the primitive to the most complex, comes with its own unique safety and health problems or hazards. Thus, the type(s) of equipment being used will determine most safety and health concerns to address in order to prevent the accidents, injuries,

Figure 1-5. Use of a powered truck for material handling and hauling

illnesses, and potential deaths associated with the particular equipment being used for material handling at each unique workplace and worksites. Many types of material handling equipment have mandatory OSHA regulations and requirement to assure equipment safety and accident prevention.

MATERIAL HANDLING REGULATIONS

The most frequently sited regulations are the General Industry (29 CFR 1910) and the Construction (29 CFR 1926) standards. Although other industry sectors have material handling issues, these general industry and construction regulations are directly representative of the requirements placed upon other industrial sectors such as Agriculture (29 CFR 1928); Ship Building (29 CFR 1916), Breaking (29 CFR 1917) and Repairing (29 CFR 1915); and Longshoring (29 CFR 1918).

General Industry

Specific regulations, applicable to material handling within General Industry (29 CFR 1910) Standards are as follows:

Subpart H – Hazardous Materials
                1910.101   – Compressed gases
                1910.102   – Acetylene
                1910.103   – Hydrogen

1910.14   – Oxygen
1910.105 – Nitrous oxide
1910.106 – Flammable and combustible liquids
1910.109 – Explosives and blasting agents
1910.110 – Storage and handling of liquefied petroleum gases
1910.107 – Storage and handling of anhydrous ammonia
Subpart J – General Environmental Controls
1910.144 – Safety color code for marking physical hazards
1910.145 – Specifications for accident prevention tags
Subpart N – Material Handling and Storage
1910.176 – Handling materials – general
1910.178 – Powered industrial trucks
1910.179 – Overhead and gantry crane
1910.180 – Crawler, locomotion, and truck cranes
1910.181 – Derricks
1910.183 – Helicopters
1910.184 – Slings
Subpart Q – Welding, cutting and brazing
1910.253 – Oxygen-fuel gas welding and cutting

Although other 29 CFR 1910 regulations are tangential, related to material handling, and apply to most workplaces, the previous list includes the ones most applicable to material handling.

Construction

Construction (29 CFR 1926) standards also have subparts and sections relevant to material handling situations. They are as follows:

Subpart C – General Safety and Health Provisions
1926.25   – Housekeeping
1926.26   – Illumination
Subpart G – Signs, Signals and Barricades
1926.200 – Accident prevention signs and tags
Subpart F – Fire Protection and Prevention
1926.152 – Flammable and combustible liquids
1926.153 – Liquid petroleum gas (LP-Gas)
Subpart H – Material Handling, Storage, Use, and Disposal
1926.250 – General requirements for storage
1926.251 – Rigging equipment for material handling
1926.252 – Disposal of waste material
Subpart J – Welding and Cutting
1926.350 – Gas welding and cutting
Subpart N – Cranes, Derricks, Hoists, Elevators, and Conveyors
1926. 550 – Cranes and derricks
1926.1927 – Helicopters
1926.1928 – Material hoist, personnel hoist, and elevators
1926.1929 – Base mounted drum hoist
1926.1930 – Overhead hoist
1926.1931 – Conveyors

Subpart O – Motor vehicles, mechanized equipment, and marine operations
            1926.600   – Equipment
            1926.601   – Motor vehicles
            1926.602   – Material handling equipment
Subpart T – Demolition
            1926.855–Manual removal of floors
Subpart U – Blasting and the Use of Explosives
Subpart V – Power Transmission and Distribution
            1926.953   – Material handling

TRAINING REQUIREMENTS

Training is often a requirement within the Occupational Safety and Health Administration's (OSHA's) safety and health regulations. At times training is mandatory and, at other times, it is recommended. There are definite requirements for all workers to be trained regarding the hazards of their particular jobs. It does little good to have trained workers if the supervisors are not trained and expected to follow and enforce the appropriate safety and health requirements and rules.

Training of workers and supervisors has been shown to have a positive effect on the reduction of accidents and incidents within the work environment. Thus, investment in training is by all means an important part of a good safety and health initiative. This is especially true when one considers the number of accidents or incidents related to material handling, which transpires and impacts upon the bottom line (profit).

Training requirements relevant to material handling can be found in Appendix A. As you can see from Appendix A, OSHA has not provided many definitive rules regarding training and material handling. The onerous responsibility is placed upon the employer to assure that his/her employees are adequately trained to perform their work in a safe and healthy manner. The newest and most specific training requirements are those for industrial truck (forklift) operators with which operators should be familiar and in compliance.

Since training cost money and time, it should be carefully planned to meet the specific material handling needs of supervisors and employees. Training should be evaluated to assure that it has accomplished its purpose. Retraining should be done when it has not been effective and new procedures arise. All training can be enhanced by designing it for the specific operation and expecting supervisors and workers to use it.

MATERIAL HANDLING AND LIFTING PROGRAM

In the field of occupational safety and health, programs and policies must be formalized for communication, compliance, and implementation to take place. In most cases, this takes the form of a written program or policy, which can be available to everyone and thus foster a better understanding of the company's, contractor's, or institution's intent and stand on pertinent issues. In light of this, it is recommended that the employers develop a written formal program to address the safety and health issues related to material handling and lifting.

Again, the employers should state their polices on safety and health practices for material handling and lifting. This should include what procedures to follow, which will vary greatly for a nursing home versus a construction site; the maximum weight an individuals is expected to lift; and guidelines on when an individual should call for help or use special devices for handling loads beyond the prescribed limit.

Second, an employer should require that workers and supervisors be trained to identify potential hazards. Procedures should be set in place for reporting hazards and for providing feedback on reported hazards. Some individuals with the appropriate expertise should be assigned responsibility for evaluating the reported hazards and responding to them.

Third, a system for addressing solutions for intervention and prevention of these hazards must exist. This could be the task of a responsible person or a team of employees and/or supervisors to find solution and put forth revise policies. A mixed team of employees and supervisor (s) will probably enact the best solutions as well as the most commitment to implementation and compliance with intervention strategies.

Fourth, employers must assure that workers receive training and training related to their material handling and lifting policies and procedures. Workers are expected to follow and know the purpose and intent of the employers' programs. Employees must be trained in material handling hazards, lifting hazards, proper material handling and lifting procedures, and safe and proper use of equipment which they must use to perform their material handling and lifting work. Supervisors should receive the same type of training. Often employers and supervisors believe they know enough, but studies often indicate the contrary. In studies, workers and their supervisors feel that additional training for supervisors has positive benefits.

Although this is only a short pointed presentation on material handling and lifting programs and policies, suffice it to say that organized approaches should be viewed as necessary rather than as luxuries in the efforts of employers, supervisors and workers to prevent the occurrence of accidents and incidents involving material handling and lifting. These programs and policies also provide a baseline for everyone in mitigating expensive types of injuries.

SAFETY AND HEALTH ISSUES

As each of us goes about our daily schedule and sets a goal to look at the various tasks, equipment, and businesses, which depend upon material handling in some form or fashion, we will notice individuals who are charged with lifting and moving materials and perhaps have a variety of other tasks to perform. What we should ask ourselves is, "Where have all of these various trucks and vehicles of all shapes, sizes, and configurations for material handling come from?" We will marvel that so many people have to understand the unique functions of material handling devices on their vehicles. As we pass warehouses, loading docks, junk yards, distribution centers, logging operations, and utility company storage areas, we will begin to understand the magnitude of ongoing material handling processes. Then come the construction sites: we can begin to count the number and types of material handling vehicles in action on the site. On many sites, the use of cranes is so extensive we might think the pet bird is the "crane." It seems that no matter in which direction we turn, there exists another example of material handling. With so much material handling going on, we can begin to realize that safety and health issues associated with material handling are unavoidable.

When we assess the risks involved in any form of work, the first item is usually exposure. Certainly, a vast majority workers have the potential for exposure. The next step is to assess the probability that exposure will result in injury or illness. Suffice it to say, the more exposure, the greater the probability that an errant event will occur. If such an event does transpire, the consequences or outcomes may range from death to injury or illness. Consequences from faulty material handling can be costly in damage, not only to human resources, but also to equipment and products.

The risk for accident and incidents related to material handling is quite high. Any initiatives, that address safe work procedures, safe equipment, and attention to material handling issues will pay large dividends in the reduction of injuries, illnesses and property damage.

PRINCIPLES OF MATERIAL HANDLING

There are some basic principles that need to be given consideration when addressing material handling issues. They are as follows in Table 1-4:

Table 1-4

Principles of Material Handling and Movement

PRINCIPLES *	EXAMPLES**/APPLICATIONS
1. **Planning Principle.** Plan all material movement and storage activities to obtain maximum efficiency and safety.	Graphic and tabular aids can be used.
2. **Systems Principle.** Integrate as many activities as is practical into a coordinated system of operations, from vendor, receiving, storage, production, inspection, packaging, warehousing, shipping, to transportation to customer.	Pallets first used for the transport of partly finished material, then through the manufacturing process, finally for shipment to the customer.
3. **Material Flow Principle.** Carefully plan process, operation sequence, and equipment, for optimizing the material flow.	Flow process chart and diagram.
4. **Simplification Principle.** Simplify material flow by eliminating, reducing, or combining movements and/or equipment.	Use one-piece die cast aluminum housing instead of milling and drilling several parts and then welding them together.
5. **Gravity Principle.** Use gravity to move material wherever practical.	Gravity feed bins, inclined roller conveyors.
6. **Space Utilization Principle.** Make optimum use of building cube.	High racks with narrow aisle, using stacking trucks, overhead conveyors.
7. **Unit Size Principle.** (a) Increase the quantity, size, or weight of unit loads so that equipment must be used for movement, or (b) reduce size and weight so that one operator can safely handle material.	(a) Pallet-size loads. (b) Reduce bulk, use light-weight material.
8. **Automation Principle.** Provide automation to include production, transports, and storage functions.	Automated stackers.
9. **Mechanization Principle.** Mechanize handling operations	Powered conveyors.
10. **Equipment Selection Principle.** In selecting equipment, consider all aspects of the material flow—the movement and the method to be used.	Fork trucks with different forks.

Table 1-4

Principles of Material Handling and Movement (Continued)

PRINCIPLES *	EXAMPLES**/APPLICATIONS
11. **Standardization Principle.** Standardized methods as well as types and sizes of equipment.	Internationally standardized cartons, pallets, containers that can be stacked.
12. **Adaptability Principle.** Use methods and equipment that can best perform a variety of tasks and applications where special purpose equipment is not justified.	Variable speed conveyors.
13. **Dead Weight Principle.** Reduce ratio of dead weight of mobile equipment to load carried.	Use of low-density materials.
14. **Utilization Principle.** Plan for optimum use of equipment and manpower.	Walkie-talkies used by truck drivers.
15. **Maintenance Principle.** Plan for preventive maintenance and scheduled repairs of all equipment.	Periodic automated function check of powered trucks.
16. **Obsolescence Principle.** Replace obsoete methods and equipment when more efficient methods or equipment will improve operations.	Look over and analyze the market for new equipment.
17. **Control Principle.** Improve control of production, inventory, and order handling.	Computerize flow control.
18. **Capacity Principle.** Use equipment to help achieve desired production capacity.	Replace manual loading/unloading of production machines by mechanized feed/removal.
19. **Performance Principle.** Determine effectiveness of performance in terms of expense per unit.	Computerized quality/quantity/cost control.
20. **Safety Principle.** Provide safe methods and equipment.	Replace manual handling by mechanized transport. Roll bars on trucks.
21. **Ergonomics Principle.** Design to best fit and use human capabilities.	Make people operators and controllers of equipment instead of having them do menial tasks, such as material handling.

* Adapted from College-Industry Committee (1966).

** Some examples adapted from Turner, Mize, and Case (1978).

MATERIAL HANDLING BENEFITS THE BOTTOM LINE

While writing this book, I visited a manufacturing operation, which had a brand new warehouse and product distribution area. As I viewed this warehouse, I realized a large amount of money had been invested. All shelving was neat and orderly, and each shelf had been bar coded for each product. All aisles had adequate space for two stock pickers to work or pass without interfering with one other.

Actually the whole warehouse and distribution process was computerized. The order comes in via computer, and the stock picker retrieves the order, which directs the driver to the location and provides the shortest path to the appropriate shelf for retrieval of the products.

If the product size is small, then that portion of the order is filled from the computerized carousel bins. The attendant or stock puller is on a computerized scissor-lift platform which moves to the appropriate carousel bin level for him or her to retrieve the product. These small products are placed in a red plastic bin that is placed upon a power driven conveyor, which transports them to a staging area where they remain until the order filler/packer calls for it. The packer presses a button to call for the next red bin, which comes automatically to the packer workstation. If the order includes larger products, the packer using a pallet truck walks some twenty feet to the staging area where the stock picker has left the pulled larger items and scanned the location into the computer for the packer. The packer transports the larger items via his/her pallet truck to a scissor-lift table, which can be progressively raised to the height of the packer work bench. The packer can complete the filling of the order without lifting any larger items. Even the packing Styrofoam peanuts are dispensed automatically into boxes from a dispensing device within easy reach above the workstation. A computerized label is printed by the computer and affixed by the packer to each box of the order. No large boxes are used. The packer slides the box(es) from his/her workstation to the conveyor that transports the box(es). The boxes are automatically weighed as well as directed to the proper roller conveyor for the shipper being used (ex. UPS). Each shipper has a specific roller convey or and staging area for the orders which they are to transport.

The conveyor tender loads the pallets for each shipper using a vacuum lift on the boxes to transfer them from the conveyor to the pallet. Once the pallet is loaded, the tender using a forklift transports the entire pallet to a turntable where the entire pallet is automatically shrink wrapped. The pallet is then transported to the automatic loading docks into waiting semi-trailers. The order filling process is completed.

Note: not one of the workers in this warehouse was working at a fever pitch or in a hurried fashion. Everyone was working at a steady pace and no one was manually lifting any heavy items: actually, very little lifting of any kind was occurring.

Then I asked the question of the general manager, "Why in the world would you make this kind of investment for a material handling process?" The answer went something like this: "I spent 50% of my time in the warehouse trying to reduce accidents, insuring that orders were accurately filled, trying to get orders out in a timely fashion, and keeping the warehouse organized."

The facts which substantiated this justification for this automated warehouse were that he could quickly see the workload as it was now arranged, the number of accidents had been reduced to zero, the previous sixty hours of overtime was now zero, and the workforce had been reduced by fourteen workers.

The day I visited the plant, the loss control specialists from his insurer were there to evaluate how the cause of previous injuries has been addressed. The company's insurance premiums for workers' compensation will be coming down. Also, the customer gets an order filled accurately and in a timely fashion.

The general manager feels that this type of safe production has a positive impact on the bottom line. The investment is worth the improved bottom line. The bottom line profit for this company is worth a safe and effective material handling process. You should consider the bottom line benefits of your own material handling activities and determine how it will not only improve your safety performance but the overall performance of your company.

INTENT AND CONTENT

It is the intent of this book to provide information on safety and health aspects of material handling. It includes applicable OSHA regulations and routine industry practices relevant to safety and health. Content covers the hazards involved and makes recommendations for the elimination or prevention of accidents and incidents associated with material handling. Also covered are topics such as manual and mechanized material handling, and descriptions of the multitude of equipment available for safe material handling.

Special emphasis is placed upon the safety issues affecting conveyors, cranes, derricks, hoist, industrial trucks (forklifts), off-highway equipment (excavators, dozers, off-road trucks, front-end loaders, and backhoes), highway vehicles (highway dump trucks and cargo transporting trucks), and railroad cars (freight, tanker, and hopper cars). Also some details of specialized equipment for handling and lifting material, including manual and powered equipment, and industrial robots, are delineated herein.

Handling and storage of materials, from solids to liquids, and the inherent hazards associated with each is presented. This includes proper stacking and the placement of hazardous and toxic chemicals into safe storage.

Safety and health are the focus throughout this text. No effort is made to discuss material handling system design, since many current books adequately address that subject. Neither does the content encompass engineering principles of material handling and storage. No effort is made to discuss in any detail the application of ergonomic or human factors associated with the handling and lifting of materials; and this book does not attempt to provide information to describe the safe movement of people (workers) via cranes, hoists, escalators, elevators, or modes of conveyance for individuals.

As can be seen, the focus of this book is on basic principles and policies for on-the-job safety and health. It has been proven that having well-established programs and policies for safety and health can help to prevent accidents and incidents, injuries, illnesses, and deaths which can arise from unsafe and unhealthy material handling, lifting, and storage. Since there are a wide variety of materials, an uncounted number of applications, and exacting, necessary procedures for material handling, lifting, and storage, my hope is that the reader will find this a useful resource and, perhaps, desk reference for keeping workers healthy and safe.

REFERENCES

United States Department of Labor. Occupational Health and Safety Administration. General Industry. *Code of Federal Regulations.* Title 29, Part 1910.Washington, GPO, 1998.

United States Department of Labor. Occupational Health and Safety Administration. Ship Repairing. *Code of Federal Regulations.* Title 29, Part 1915.Washington, GPO, 1998.

United States Department of Labor. Occupational Health and Safety Administration. Ship Building. *Code of Federal Regulations.* Title 29, Part 1916.Washington, GPO, 1998.

United States Department of Labor. Occupational Health and Safety Administration. Ship Breaking. *Code of Federal Regulations.* Title 29, Part 1917.Washington, GPO, 1998.

United States Department of Labor. Occupational Health and Safety Administration. Longshoring. *Code of Federal Regulations*. Title 29, Part 1918.Washington, GPO, 1998.

United States Department of Labor. Occupational Health and Safety Administration. Construction. *Code of Federal Regulations*. Title 29, Part 1926.Washington, GPO, 1998.

United States Department of Labor. Occupational Health and Safety Administration. Agriculture. *Code of Federal Regulations*. Title 29, Part 1928.Washington, GPO, 1998.

CHAPTER 2

MANUAL MATERIAL HANDLING

Construction worker manually handling rebar

Manual material handling by definition refers to a workers using their hands to bodily lift and handle materials. It seems appropriate to expand the discussion beyond simply the worker's physical effort to the use of devices, which make the manual effort more bearable. Although aids to lifting are not what we would consider to be powered lifting devices, they definitely increase the ability of the worker to handle loads in a safer manner. Due to the leverage provided by these lifting aids, the worker is able to move material that which would be considered too heavy to move or lift without devices. These types of lifting devices seldom remove the manual part of material handling activity completely.

Overexertion causes 28 percent (NSC Accident Facts, 1998) of all reported injuries, the bulk of which are due to moving, handling and/or lifting of something within a work environment. This suggests strongly that more careful attention needs to be paid to the manual side of material handling.

Certainly workers who do manual lifting can suffer injuries such as cuts, crushed fingers, and toes, contusions, bruises, or fractures, but the most common manual material handling injury are sprains and strains: for example, sprains and strains of the back are the most common part of the body affected. These types of injuries are most often classified as overexertion injuries.

Although anything that can be done to decrease the number of back injuries is good, it is unrealistic to expect 100 percent prevention. However, the reduction in frequency and severity is a more realistic goal.

Workers frequently cite the weight and bulkiness of objects being lifted as major contributing factors to their injuries, according to the National Safety Council's Accident Facts (1993), and one-third of all injuries are back injuries. The second factor frequently cited by workers as contributing to their injuries was bodily movement. Bending, followed by twisting and turning, was the most commonly cited movement that caused back injuries.

MANUALLY LIFTING AND HANDLING MATERIALS

The use of muscle power has been a primary mover in the handling of materials since time one. Although a great deal of manual handling continues today, workers are probably not as prepared as well mentally and physically to do manual tasks. Often companies hire the very young to do the heavy lifting tasks, since they tend to be in better physical condition and have less wear and tear upon their bodies. Our society today does not usually expect workers to be conditioned enough to carry out heavy lifting tasks; thus, the industries are leaning heavily towards mechanizing their lifting tasks as much as possible. Also, it certainly appears that the average worker today is older and frequently less capability of performing heavy lifting tasks. Some workers are required to lift heavy and awkward items, such as household movers, package handling and sorting workers, construction workers, and restocking workers. As long as materials are handled manually, it is expected that injuries will occur because of the load's weight, its location, or because of the number of repetitious movements while lifting.

BACK INJURIES

No matter what is written, researched, and implemented regarding back injuries, back injuries are pervasive and seem to be endemic within the U.S. workforce, and some might say epidemic in proportion. All the emphasis programs, training, gimmicks, and gadgets have not been able to restrict the continuous flood of back injuries. They have only seemed to be stop-gap procedures.

Suffice it to say, at this time back injuries are not entirely preventable when a worker must manually lift heavy loads, move loads some distance, or perform repetitiously lifting and moving of materials. Certainly, workers and supervisors must be trained in proper lifting procedures, and these lifting procedures need to be constantly reinforced so that workers are encouraged to avoid deviations considered inconvenient or interfering with production.

Thus, employers must do more than train workers. A lifting policy must be in place, to limit the maximum weight a worker should lift. It should delineate procedures workers should use when evaluating new or infrequent lifts. Workers should know that they have the latitude to request assistance or procure the appropriate lifting aid. Lifting should never be put forth as a demonstration or contest of strength.

To decrease back injuries, three specific items must be accomplished:

1. Decrease the weight of materials to be lifted.
2. Decrease the distance the materials must be moved.
3. Decrease the number of lifts or repetitions, which must be done.

Of course, any lifting activity that can be mechanized or automated should be undertaken in order to decrease back injuries. But, remember, any time a new procedure or piece of equipment is placed into service, other hazards are likely to arise if those involved are not safeguarded and properly trained on the new procedure or equipment. It cannot be assumed that everyone is on the same page of the playbook.

With this said, the rest of this chapter presents back injury prevention. It includes a mix of traditional wisdom and non-traditional information relevant to back injury prevention. The most commonly listed causes of back injuries are:

- Failure to use proper lifting techniques that avoid twisting and jerking motions.
- Repeated strenuous lifting.
- Inadequate evaluation of the weight of the material being handled.
- Poor communications between persons lifting and carrying loads.
- Not using readily available means of handling material such as a hand truck, fork-lift, or front-end loader.

To review, guidelines for a safe lift adhere to the following. (See Figure 2-1.)

1. Plan ahead what you want to lift and do not be in a hurry.
2. Separate your feet to shoulder-width to give yourself a solid base of support.
3. Bend your knees.
4. Grasp using the palm grip, which is stronger than using just fingers.
5. Tighten your stomach muscles.
6. Position the object to be lifted close to your body.
7. Lift with your legs.
8. Avoid twisting your body.
9. Point your toes in the direction you want to move and pivot in that direction.
10. Do not try to lift an object that is too heavy or an awkward shape. Get help.
11. Maintain the natural curve of your spine; don't bend at your waist.
12. Do get help when lifting something too heavy for you.

A As you approach the load determine its weight, size and shape. Consider your physical ability.

B Stand close to the object with feet 8 to 12 inches apart for good balance.

C Bend the knees and get a firm grasp on the object.

D Using both leg and back muscles lift the load straight up. Keep the object close to the body.

E Do not twist or turn until the object is in carrying position.

F Rotate body by turning your feet and make sure path of travel is clear.

G To set the object down, use leg and back muscles and lower the object by bending the knees.

Figure 2-1. Lifting using safe lifting practices. Courtesy of the Bureau of Mines

Even when a worker lifts properly, there are other risk factors, which come into play. The knowledge relevant to risk factors is always changing but some key predictors of work-related mishaps are listed on the following instrument. Any item checked is an indicator of potential risk for back injuries. (See Figure 2-2.)

RISK FACTOR INSTRUMENT

Check all of the following which applies to you during your work activities.

_____ Your job requires heavy lifting (weights greater than 50 pounds).

_____ You are required to perform lifting tasks in a hurried fashion.

_____ You are overweight.

_____ You do not exercise regularly.

_____ You are required to stretch, reach, or twist your body when lifting or moving objects.

_____ Your job requires frequent lifting.

_____ You are a smoker.

_____ Your job requires you to keep your legs stiff when lifting.

_____ Your job requires frequent bending of your back in order to lift.

_____ You have a slouching posture when you stand or sit at your job.

A check mark on any of the above statements indicates that you are at risk or have the potential to experience a back injury.

Figure 2-2. Risk factor instrument

When workers are required to perform a new lifting task, they should take time to evaluate the lift, make sure other hazards are removed prior to the lift, and listen to their body to assure that they can make the lift safely. To simplify this approach further, **Stop, Look, and Listen!**

Many documents and books discuss in some detail the NIOSH lifting guidelines and calculations, which seem logical and are backed by the state-of-the-art science and research. The original lifting document was for symmetrical lifting. Many of the lifting tasks in the workplace are asymmetrical in nature. The revised lifting formula has incorporated the asymmetrical component to address the previous deficiency.

The following is a summary of the intent and formula found in the *Applications Manual for the Revised NIOSH Lifting Equation*. The revised NIOSH equation provides a set of evaluation parameters that include: (1) intermediate task-related multipliers to define the extent of physical stress associated with individual task factors; (2) NIOSH's Recommended Weight Limit (RWL), a value that defines the weight load for specific tasks, which are reasonably safe for most healthy workers; and (3) the NIOSH Lifting Index (LI), which provides an estimate of the overall physical stress associated with a specific manual lifting task.

The RWL is defined as follows.

$$RWL = LC \times HM \times VM \times DM \times AM \times CM \times FM$$

The load constant (LC) is equal to 51 pounds and the remainder of the equation consist of task-specific multipliers which tend to reduce the recommended weight limit (RWL) according to the specified task factor to which each multiplier applies (see Table 2-1). The magnitude of each multiplier will range in value between zero and one, depending on the value of the task factor to which the multiplier applies. The multipliers are defined as follows in Table 2-2 and 2-3. A sample lifting analysis worksheet is depicted in Figure 2-3.

Table 2-1

Task Variables

Component	Metric	U.S. Customary
LC = Load constant	23 kg	51 lbs.
HM = horizontal multiplier	(25/H)	(10/H)
VM = vertical multiplier	$1-(.003\,|\,V-75\,|)$	$1-(.0075|\,V-30\,|)$
DM = distance multiplier	.82 + (4.5/D)	.82 + (1.8/D)
AM = asymmetric multiplier	$1-(.0032A)$	$1-(.0032A)$
FM = frequency multiplier	(determined from Table 2-2)	
CM = coupling multiplier	(determined from Table 2-3)	

Source: *Applications Manual for the Revised NIOSH Lifting Equation.*

Table 2-2

Frequency Multiplier (FM)

Frequency Lifts/min	Work Duration					
	< 1 Hour		< 2 Hours		< 8 Hours	
	V < 75	V > 75	V < 75	V > 75	V < 75	V > 75
0.2	1.00	1.00	.95	.95	.85	.85
0.5	.97	.97	.92	.92	.81	.81
1	.94	.94	.88	.88	.75	.75
2	.91	.91	.84	.84	.65	.65
3	.88	.88	.79	.79	.55	.55
4	.84	.84	.72	.72	.45	.45
5	.80	.80	.60	.60	.35	.35
6	.75	.75	.50	.50	.27	.27
7	.70	.70	.42	.42	.22	.22
8	.60	.60	.35	.35	.18	.18
9	.52	.52	.30	.30	.00	.15
10	.45	.45	.26	.26	.00	.13
11	.41	.41	.00	.23	.00	.00
12	.37	.37	.00	.21	.00	.00
13	.00	.34	.00	.00	.00	.00
14	.00	.31	.00	.00	.00	.00
15	.00	.28	.00	.00	.00	.00
>15	.00	.00	.00	.00	.00	.00

Values of V are in cm; 75 cm = 30 in.

Source: *Applications Manual for the Revised NIOSH Lifting Equation.*

Table 2 -3

Coupling Multiplier (CM)

Couplings	V < 75 cm (30 in.)	V > 75 cm (30 in.)
Good	1.00	1.00
Fair	0.95	1.00
Poor	0.90	0.90

Source: *Applications Manual for the Revised NIOSH Lifting Equation.*

JOB ANALYSIS WORKSHEET

DEPARTMENT _____   **JOB DESCRIPTION**
JOB TITLE _____   _____
ANALYST'S NAME _____   _____
DATE _____   _____

STEP 1. Measure and record task variables

Object Weight (lbs)		Hand Location (in)				Vertical Distance (in)	Asymmetric Angle (degrees)		Frequency Rate	Duration	Object Coupling
		Origin		Dest			Origin	Destination	lifts/min	(HRS)	
L (AVG)	L (Max)	H	V	H	V	D	A	A	F		C

STEP 2. Determine the multipliers and compute the RWL's

RWL = LC · HM · VM · DM · AM · FM · CM

ORIGIN RWL = 51 · ☐ · ☐ · ☐ · ☐ · ☐ · ☐ = ☐ Lbs

DESTINATION RWL = 51 · ☐ · ☐ · ☐ · ☐ · ☐ · ☐ = ☐ Lbs

STEP 3. Compute the LIFTING INDEX

ORIGIN LIFTING INDEX = $\dfrac{\text{OBJECT WEIGHT (L)}}{\text{RWL}}$ = ———— = ☐

DESTINATION LIFTING INDEX = $\dfrac{\text{OBJECT WEIGHT (L)}}{\text{RWL}}$ = ———— = ☐

Figure 2-3. Job analysis worksheet. Courtesy of the National Institute for Occupational Safety and Health

In order to use the equation, the analyst must make the following measurements:

L = Weight of the load being lifted (kg or lbs.).
H = Horizontal location of hands from midpoint between the ankles. Measure at the origin and the destination of the lift (cm or in.).
V = Vertical location of the hands from the floor. Measure at the origin and destination of the lift (cm or in).
D = Vertical travel distance between the origin and the destination of the lift (cm or in.).
A = Angle of asymmetry is defined as the angular displacement of the load from the sagittal plane when lifts are made to the side of the body. Measure at the origin and destination of the lift (degrees).
F = Average frequency rate of lifting measured in lifts/min.

Duration is defined to be: < 1 hour; < 2 hours; or < 8 hours assuming appropriate recovery allowances. (See Table 2 - 2.)

The Lifting Index (LI) provides a relative estimate of the level of physical stress associated with a particular manual lifting task. The estimate of the level of physical stress is calculated by the relationship of the weight of the load lifted and the recommended weight limit. The formula for LI is as follows:

$$LI = \frac{\text{Load Weight (L)}}{\text{Recommended Weight Limit (RWL)}}$$

A detailed explanation of the use of the revised NIOSH lifting equation, including definitions of terms and procedures, is available in the *Applications Manual for the Revised NIOSH Lifting Equation* (Waters, 1994).

PERSONAL PROTECTIVE EQUIPMENT

It seems that it is frequently a battle to get workers to wear personal protective equipment (PPE). When the cost of one eye injury is usually more than the cost of providing eye protection for a year, it seems to be a logical conclusion that to enforce the wearing of personal protective equipment makes sense. The requirement to conduct a hazard analysis as required by 29 CFR 1910.132, the Personal Protective Equipment regulations, will allow for the identification of the appropriate PPE needed in that workplace. A personal protective equipment hazard assessment certification form can be found in Appendix B to use for this purpose.

Although not required by regulations for all workplaces, a strong recommendation would be that at a minimum, workers should be required to wear protective eye wear with side shields, and hand protection (gloves) during material handling activities. There is no logical reason for workers not to protect the eyes and hands during material handling activities. Also, after a hazard assessment, policies may be set in place to require protective foot wear, hard hats, chemical protective equipment, or padded protective devices.

Flying objects instantaneously soar at speeds beyond human reaction times. Since manufacturers produce a wide variety of stylish, lightweight, safety eyewear, comfortable protective eyewear can be found for each and every worker. Thus, requiring protective eyewear for workers and supervisors will not be a hardship on them.

Since hand protection is an important aspect of material handling activities, routine use of a good pair of leather gloves will provide some of the best protection from cuts, lacerations, low voltage electrical shocks, and vibration injuries. Gloves should be worn when handling materials that have sharp edges and when climbing. Other types of gloves that may be needed are cut-resistant gloves and gloves with gripping materials on the palms and fingers. Some material handling applications may also require the use of special-made gloves or chemical resistant gloves. Gloves seldom prevent punctures or crushing injuries but may decrease the severity. The most common causes of hand injuries are:

- Wearing jewelry and/or loose fitting clothing.
- Not wearing personal protective equipment.
- Failure to concentrate on the task being performed.
- Not inspecting material for sharp or jagged edges and heat before handling.
- Failure to avoid pinch points on rotating and/or moving machine parts.

Hard hats should be used whenever the slightest possibility of material striking or falling on a worker's head exists. The use of a chinstrap has not become popular in the U.S. workforce but is required when working around helicopters. Wearing head protection is truly not a difficult policy or rule to implement or enforce.

The requirement for hard-toed safety shoes is necessary when there is potential for material to drop on or roll over workers' feet. Stories have been told of the steel toes being crushed when objects fall on or roll over a worker's foot, trapping or severing his/her toes. This story is told often with no verification or documentation. Realistically if something is heavy enough to crush the steel-toed shoes' boxes, the worker would in all likelihood not have any toes left if he/she were not wearing steel-toed safety shoes. It is reasonable to require steel-toed shoes to prevent foot injuries. If more protection is needed, then the use of metatarsal protection can be required.

In foot protection, there have been advancements in the composition of the soles for working surfaces. In order to prevent slips, trips, or falls, workers may need soles with aluminum oxide embedded to provide better traction. When handling and moving materials, better traction may help prevent the most common types of accidents experienced by all industries slips, trips, and falls.

Depending upon the work surfaces evaluation, the most appropriate soles for existing work surfaces can be procured.

Other types of protective wear, such as forearm protection, may be necessary when an assessment is made of the types of accidents and body parts most frequently injured. Selection of other PPE enforcement of and compliance with company's or contractor's policies to wear protective gear can be an investment in a more profitable bottom line due to decreased injuries and reduced workers' compensation premiums.

LIFTING AND MATERIAL HANDLING AIDS

Anytime any kind of lifting aid can be put to use, it is easier than manually handling the existing load. Lifting aids may be the attachment of temporary handles, the use of straps, or the use of wheeled trucks, dollies, or carts. Whenever handles, hand holds, or cutouts can be incorporated as a lifting aid, the handling of the load is easier. None of these examples of lifting aids are powered by anything except human strength. But all provide a degree of mechanical advantage, which allows for easier handling, easier movement, and less force to accomplish the same task. (See Figures 2-4 and 2-5.)

No matter what lifting device or aid is selected to use, it should be inspected prior to use and removed from service if defects exist. If the device is determined to be safe, then an assessment of its appropriateness for the manual material handling task needs to be made. Lifting aids must be capable of handling the weight of the load. This is especially true for any type of crane, hoist, or derrick. If the operator or supervisor will take time to evaluate the material handling task, he or she will most certainly discover that there are material handling or lifting aids that can be adapted to the specific requirements. In addition, these devices are generally not as expensive as one would presuppose them to be.

Figure 2-4. Pallet truck

Figure 2-5. Manual handling aid (wheelbarrow)

WALKING AND WORKING SURFACES

If material must be handled, moved, or lifted, the foundation upon which this is to be achieved is a critical component in accomplishing material handling. How solid is the foundation? Is it concrete or soil? Is it smooth or rough? Does it have traction? Is it slick? Is it uneven? Is there an incline? Are there distortions? Is it cluttered? Do tripping hazards exist? Are there adverse weather conditions?

All of these are legitimate questions regarding walking and working surfaces. Depending upon the nature and magnitude of the material handling task, many problems regarding the surface may need to be addressed. Solutions might be as drastic as resurfacing an area to as simple as performing housekeeping. But, proper foot wear should never be overlooked when materials are to be handled or moved. Since slips, trips, and falls are among the most common types of accidents, the walking and working surface can be one of the foremost contributors to their occurrence.

PLANNING MANUAL MATERIAL HANDLING

Quite possibly a material handling checklist devised by NIOSH (see Figure 2-6) can help and be put to use in assessing a material handling problems. This checklist will help determine where possible problem areas are. After having done a preliminary assessment, the need may exist to redesign either the lifting and lowering task, pushing and pulling task, or carrying task. Figures 2-7, 2-8, and 2-9 from the NIOSH's *Elements of Ergonomics Programs Primer* should be helpful in altering the design of material handling activities.

Tray 5–F. Materials Handling Checklist

"No" responses indicate potential problem areas which should receive further investigation.

1.	Are the weights of loads to be lifted judged acceptable by the workforce?	❑ yes	❑ no
2.	Are materials moved over minimum distances?	❑ yes	❑ no
3.	Is the distance between the object load and the body minimized?	❑ yes	❑ no
4.	Are walking surfaces		
	level?	❑ yes	❑ no
	wide enough?	❑ yes	❑ no
	clean and dry?	❑ yes	❑ no
5.	Are objects		
	easy to grasp?	❑ yes	❑ no
	stable?	❑ yes	❑ no
	able to be held without slipping?	❑ yes	❑ no
6.	Are there handholds on these objects?	❑ yes	❑ no
7.	When required, do gloves fit properly?	❑ yes	❑ no
8.	Is the proper footwear worn?	❑ yes	❑ no
9.	Is there enough room to maneuver?	❑ yes	❑ no
10.	Are mechanical aids used whenever possible?	❑ yes	❑ no
11.	Are working surfaces adjustable to the best handling heights?	❑ yes	❑ no
12.	Does material handling avoid		
	movements below knuckle height and above shoulder height?	❑ yes	❑ no
	static muscle loading?	❑ yes	❑ no
	sudden movements during handling?	❑ yes	❑ no
	twisting at the waist?	❑ yes	❑ no
	extended reaching?	❑ yes	❑ no
13.	Is help available for heavy or awkward lifts?	❑ yes	❑ no
14.	Are high rates of repetition avoided by		
	job rotation?	❑ yes	❑ no
	self-pacing?	❑ yes	❑ no
	sufficient pauses?	❑ yes	❑ no
15.	Are pushing or pulling forces reduced or eliminated?	❑ yes	❑ no
16.	Does the employee have an unobstructed view of handling the task?	❑ yes	❑ no
17.	Is there a preventive maintenance program for equipment?	❑ yes	❑ no
18.	Are workers trained in correct handling and lifting procedures?	❑ yes	❑ no

---

Fig. 2-6. Material handling checklist. Courtesy of the National Institute for Occupational Safety and Health

Tray 9–D. Design Principles for Lifting and Lowering Tasks

1. Optimize material flow through the workplace by

    — reducing manual lifting of materials to a minimum,
    — establishing adequate receiving, storage, and shipping facilities, and
    — maintaining adequate clearances in aisle and access areas.

2. Eliminate the need to lift or lower manually by

    — increasing the weight to a point where it must be mechanically handled,
    — palletizing handling of raw materials and products, and
    — using unit load concept (bulk handling in large bins or containers).

3. Reduce the weight of the object by

    — reducing the weight and capacity of the container,
    — reducing the load in the container, and
    — limiting the quantity per container to suppliers.

4. Reduce the hand distance from the body by

    — changing the shape of the object or container so that it can be held closer to the body, and
    — providing grips or handles for enabling the load to be held closer to the body.

5. Convert load lifting, carrying, and lowering movements to a push or pull by providing

    — conveyors,
    — ball caster tables,
    — hand trucks, and
    — four-wheel carts.

*Adapted from design checklists developed by Dave Ridyard, CPE, CIH, CSP. Applied Ergonomics Technology, 270 Mather Road, Jenkintown, PA 19046–3129.

Figure 2- 7. Lifting and lowering tasks checklist. Courtesy of the National Institute for Occupational Safety and Health

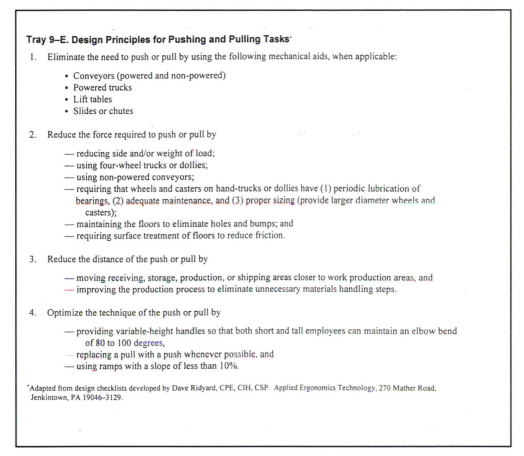

Tray 9–E. Design Principles for Pushing and Pulling Tasks

1. Eliminate the need to push or pull by using the following mechanical aids, when applicable:

    - Conveyors (powered and non-powered)
    - Powered trucks
    - Lift tables
    - Slides or chutes

2. Reduce the force required to push or pull by

    — reducing side and/or weight of load;
    — using four-wheel trucks or dollies;
    — using non-powered conveyors;
    — requiring that wheels and casters on hand-trucks or dollies have (1) periodic lubrication of bearings, (2) adequate maintenance, and (3) proper sizing (provide larger diameter wheels and casters);
    — maintaining the floors to eliminate holes and bumps; and
    — requiring surface treatment of floors to reduce friction.

3. Reduce the distance of the push or pull by

    — moving receiving, storage, production, or shipping areas closer to work production areas, and
    — improving the production process to eliminate unnecessary materials handling steps.

4. Optimize the technique of the push or pull by

    — providing variable-height handles so that both short and tall employees can maintain an elbow bend of 80 to 100 degrees,
    — replacing a pull with a push whenever possible, and
    — using ramps with a slope of less than 10%.

*Adapted from design checklists developed by Dave Ridyard, CPE, CIH, CSP. Applied Ergonomics Technology, 270 Mather Road, Jenkintown, PA 19046–3129.

Figure 2-8. Pushing and pulling tasks checklist. Courtesy of the National Institute for Occupational Safety and Health

Tray 9–F. Design Principles for Carrying Tasks

1. Eliminate the need to carry by rearranging the workplace to eliminate unnecessary materials movement and using the following mechanical handling aids, when applicable:

   - Conveyors (all kinds)
   - Lift trucks and hand trucks
   - Tables or slides between workstations
   - Four-wheel carts or dollies
   - Air or gravity press ejection systems

2. Reduce the weight that is carried by

   — reducing the weight of the object,
   — reducing the weight of the container,
   — reducing the load in the container, and
   — reducing the quantity per container to suppliers.

3. Reduce the bulk of the materials that are carried by

   — reducing the size or shape of the object or container,
   — providing handles or hand-grips that allow materials to be held close to the body, and
   — assigning the job to two or more persons.

4. Reduce the carrying distance by

   — moving receiving, storage, or shipping areas closer to production areas, and
   — using powered and nonpowered conveyors.

5. Convert carry to push or pull by

   — using nonpowered conveyors, and
   — using hand trucks and push carts.

*Adapted from design checklists developed by Dave Ridyard, CPE, CIH, CSP. Applied Ergonomics Technology, 270 Mather Road, Jenkintown, PA 19046–3129.

Figure 2-9. Carrying tasks checklist. Courtesy of the National Institute for Occupational Safety and Health

REFERENCES

Cohen, A. L., et al., *Elements of Ergonomics Programs*. U.S. Department of Health and Human Services/NIOSH, Cincinnati, OH, 1997.

National Safety Council. Accident Facts. Itasca, IL, 1993 and 1998.

United States Department of Interior. Bureau of Mines., *Technology Transfer Symposium Proceedings*–Back Injuries. Pittsburgh, PA, 1983.

Waters, T.R., Putz-Anderson, V. and Garg, A, *Applications Manual for the Revised NIOSH Lifting Equation*. United States Department of Health and Human Services/ NIOSH, Cincinnati, OH, January 1994.

CHAPTER 3

POWER-ASSISTED MATERIAL HANDLING

Piggy back handling works well for these trucks

In most cases, anytime a manual (human powered) material handling task can be augmented or replaced by a powered piece of lifting or handling equipment, the potential from overexertion injuries to workers will be greatly reduced. This type of equipment decreases the weight the worker must handle, the distance the loads must be moved, and the number of repetitious lifts.

Powered equipment allows for the handling of loads too heavy for workers to manually handle. This type of equipment often decreases the time taken to lift, handle, and move loads. The variety of power-assisted equipment available suggests that, with some research and problem solving skills, powered equipment that meets existing material handling needs can be found. Power-assisted tools provide a mechanical advantage for lifting, positioning, and moving materials.

Powered lifting equipment, such as pallet trucks, hoists, material lifts, cranes, stackers, lift/tilt tables, powered conveyors, and powered trucks, has many applications within the material handling arena. When using power-assisted material handling equipment, these specific rules should be followed:

1. Insure that the equipment's load rating is adequate to handle the load.

2. Make sure that the load is centered and balanced to prevent shifting or tilting of equipment.

3. Make sure the load is properly secured to prevent shifting or falling of the load.

4. Insure that the operators are trained to operate the specific equipment safely and are aware of the equipment's limitations and hazards.

5. Caution workers about the potential of loads to shift or fall. Workers should never be under loads unless they are secured, supported, or blocked to prevent falling.

6. Inspect all equipment daily and repair or remove it from service if defects are found prior to use. Equipment is not to be used until free of defects.

7. Maintenance should be a standard part of the safe operating procedures when power-assisted material handling equipment is used.

8. Make certain the employers and workers are using the appropriate power-assisted material handling equipment which was designed to perform the task at hand.

Almost any material handling task can be mechanized. Many innovations have transpired in the past decades. Today, for example, moving storage, shelves are designed to move up and down and in and out to allow for easier access. Most material handling equipment comes with attachments, or attachments can be purchased, which allow the equipment to be used for different tasks. The versatile forklift comes with attachments, which allow it to be converted to a stock picker or drum carrier. Electric vehicles (trucks) can transport supplies, tools, and parts and have found their way into large/spread out-workplaces. (See Figure 3-1.)

Figure 3-1. Industrial vehicles used for equipment and people transport

Hoists, vehicles, and cranes provide power assistance via hydraulic or electrical power and are in wide use when material handling tasks warrant it. (See Figure 3-2.)

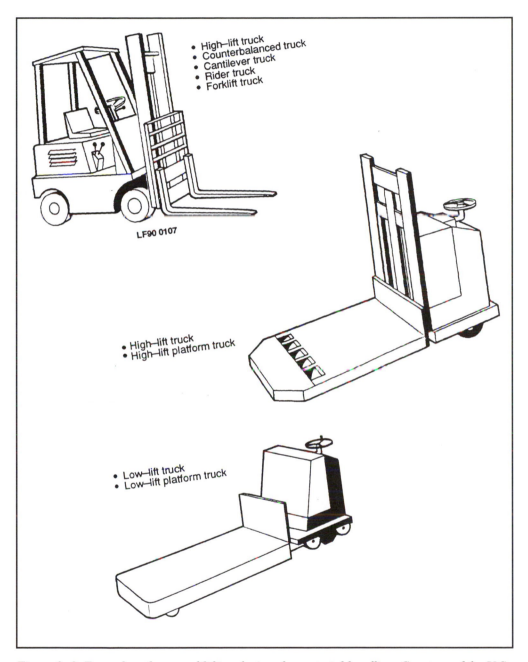

Figure 3- 2. Examples of powered lifting devices for material handling. Courtesy of the U.S. Department of Energy

Certainly, power-assisted material handling equipment can perform jobs beyond the capacity of worker. With this power come powerful sources of energy, which, if not controlled, can cause damage, accidents, and/or injuries. While convenient, powered equipment comes with unique hazards, which are now part of our work environment.

Some general guidelines should be followed for working with powered equipment are the following:

1. Make sure all rotating or energized parts are properly guarded.

2. Make sure equipment is inspected for defects or malfunction prior to use.

3. Make sure all switches are operational.

4. Make sure the equipment has the capacity to handle the load.

5. Make sure the equipment is proper for the material handling task.

6. Make sure workers are properly trained in the safe operation of the equipment.

7. Make sure that workers understand the proper handling procedures for the hazards involved with the fuels (gasoline or propane), electricity (battery), or pneumatic.

8. Make sure no maintenance is performed unless the equipment has been deenergized or locked out/tagged out.

9. Make sure proper preventive maintenance is performed.

10. Make sure unqualified individuals do not operate the equipment.

11. Make sure that all those operating powered equipment are qualified operators.

The use of powered equipment may be for lifting, transporting materials, positioning material, or placing materials. An example of powered lifting equipment can be found in Figure 3-3 in the form of a powered crane.

Figure 3- 3. Powered truck crane

The transportation of materials often requires equipment with special adaptation or attachments. An example of powered equipment moving materials is found in Figure 3-4 in the form of an end loader with an attachment for handling of rebar on a construction site.

Figure 3- 4. Frontend loader with attachment for handling rebar with a sling

The positioning of materials should make them more efficiently accessible to workers while reducing the hazards of handling material in precarious positions. Figure 3-5 shows the use of a lift table to position materials for easier access by workers. Many other adaptations or pieces of equipment can be used to position materials for workers.

Figure 3- 5. Worker using a lift table for easier access

The placement of materials from one area to another is often done by powered industrial mobile lifting equipment such as the reach truck in Figure 3-6. Many variations of these types of powered equipment can be found in the industrial setting today. Figure 3-7 shows a variety of these types of powered equipment.

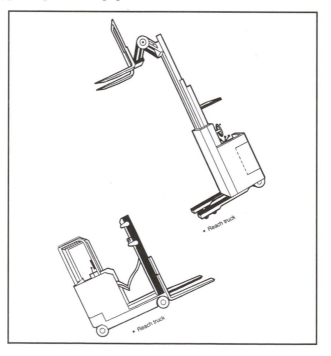

Figure 3- 6. Powered reach trucks. Courtesy of the U.S. Department of Energy

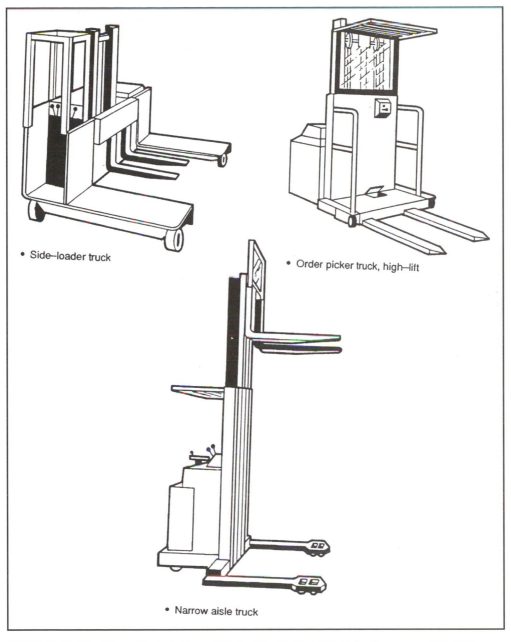

- Side–loader truck
- Order picker truck, high–lift
- Narrow aisle truck

*Figure 3- 7. Examples of powered equipment for placing materials. Courtesy of the U.S.
Department of Energy*

THE OPERATOR

Although emphasized previously, it certainly makes sense that well-trained and qualified operators on the specific piece of equipment or vehicle are imperative to the safe handling,
lifting, and transport of materials within the U.S. workplace. Only qualified operators should
be permitted to operate the equipment.

The use of cranes, forklifts, hoists, in-plant powered industrial trucks, and other powered material handling equipment are subject to certain hazards that cannot be met by mechanical means. Only by the exercise of intelligence, care, and good sense can these hazards be met. It is essential to have competent and careful operators who are physically and mentally fit and thoroughly trained regarding the safe operation of the equipment and the handling of loads. Serious hazards include the overloading, dropping or slipping of the load caused by improper hitching or slinging; obstruction to the passage of the load; or using equipment for a purpose for which it was not intended or designed.

It is necessary that persons who operate powered equipment to learn and understand the basic information concerning safety for that equipment. Also, it is necessary that these operators learn the special requirements for safe handling and use of the equipment. Since each item of equipment has its own special requirements, employees need additional training to the basic information they may have received applicable to a general class of equipment. The scope of information given here covers the general needs of an equipment class. Special requirements on specific equipment will require more training on each machine. Training for equipment operation requires two parts:

1. An information exchange when rules, regulations, requirements, limits, and dos and don'ts are discussed and explained.

2. The physical application where safe operation is explained, demonstrated by the teacher, tried by the trainee, faults and errors corrected, and checks made during operation to assure that correct physical controls are developed.

An employee cannot be a good operator until both parts are learned.

Operators of powered material handling equipment should meet the following requirements:

1. Age – At least 18 years.

2. Language – Understand spoken and written English, or a language generally in use at the location.

3. Physical – Meet the physical requirements.

4. Knowledge – Have basic knowledge and understanding of equipment-operating characteristics, capabilities, and limitations including: equipment rated capacity and effect of variables on capacity, safety features, required operating procedures, and requirements.

5. Skill – Must demonstrate skill in manipulations and control of equipment through all phases of operation.

The initial qualification of operators shall include:

1. Training on the equipment for which the operator is to be qualified, under the direction of a qualified operator designated by management to instruct in the operation of hoisting equipment.

2. Reviewing the worker's knowledge, including written and oral examinations, and witnessing a demonstration of his/her skills by the instructor.

3. Inserting a written record of training, competency, and authorization in the employee's training record (by his/her supervisor). The record shall include identification of the equipment for which the individual is qualified to operate.

4. Operator qualification is for a period of usually three years, unless the qualification is revoked sooner by the operator's manager. If operators are disqualified, their manager should enter the action in their files.

Training Programs

All organizations employing personnel who operate powered material handling equipment should develop training programs, including a means of testing, to assure that personnel are competent to perform the operations. The people responsible for safety within the organization must review program content for safety significance and include in their routine audits the administration of the compliance with the training and qualification program established and approved by the cognizant manager. Training programs for operators should address two levels of required performance:

1   Persons who may operate powered material handling equipment as an incidental part of their normal work assignment.

2.   Persons whose principal assignment is the operation of the powered equipment.

Training programs should include, but not be limited to, written tests, field training and trials, personal physical requirements and examinations, trainee status and training procedures, high consequence operations training and briefing, and qualification authority. Other topics that should be covered are:

- Travel.
- Access and egress during normal and emergency conditions.
- Preshift check procedures, including:
- Power.
- Equipment condition.
- Controls.
- Limit switch.
- Brakes.
- Fire extinguisher, signal horns, bells, etc.
- Handling of hoisting mechanism.
- Knowledge of hazards.
- Operation of controls and test lift.
- Signal reception both hand and sound.
- Operating procedures and safe practices.
- Knowledge of terminology as applicable to the equipment to be operated.
- Proper conduct.
- Special requirements for OSHA.
- Proper shutdown.
- Release energy and lockout for maintenance and overhaul.

Powered material handling equipment requires not only careful attention to the equipment's capability and operational capacity, but careful attention to the operators who are to run it. With powered material handling equipment the safety of the equipment, the safe operation of procedures, and the safe operation by the worker are of primary concern to those who must work with and around such equipment.

REFERENCES

United States Department of Energy. *Hoisting and Rigging Manual.* Washington, DC, 1991.

United State Department of Labor. OSHA Office of Training and Education. *OSHA Voluntary Compliance Outreach Program: Instructors Reference Manual.* Des Plaines, IL, 1993.

CHAPTER 4

CONVEYORS

Use of multiple conveyors at a sand and gravel surface mine

Conveyors in the workplace can be of two types: those which are essentially gravity or inertia operated, and those which are operated as motorized belts. Of course, both have their unique problems, which is why each type will be considered separately. When conveyors are used, workers' hands may be caught in nip points, material may fall onto workers from the conveyor, or material may fly off conveyors, especially ones traveling at excessive speeds, causing the material to strike a worker.

GRAVITY OR INERTIA CONVEYORS

Gravity or inertia conveyors may pose a myriad of problems for the users. First, material handling processes may require workers to lift or raise materials to the height of the conveyor. Even if this is not the case, workers are required to repetitiously place materials or loads upon the conveyor. To do so may result in having to twist the body. Along with this, the worker may be required to use a great deal of force to propel the materials on their way down the conveying system. At times materials may be traveling very rapidly. At the end destination the worker may use manual force to slow or stop the rate of its travel on the conveyor. Even if these are not issues, conveyors have a way of increasing the speed of a task, whether it be loading or unloading of the conveyor. Thus, due to the expedited work speed, workers are again placed in a stressful situation.

Although inherent hazards exist, it is no doubt better to move materials via a conveyor from one point to another rather than carrying the material or load. Conveyors can increase the speed of accomplishing a task. The main types of gravity or inertia conveyors are the skate and the roller. (See Figure 4-1 and 4-2.)

Figure 4-1. Gravity or inertia conveyor (skate type)

Although workers will usually be required to move items onto the conveyor, the use of adjustable lift tables or adjustable workstations that are the same height as the conveyors can facilitate the loading and unloading of these material movers. When workers load and unload these types of conveyors, whether they are the skate wheel or roller type, pinch points or sharp edges may exist. Also, the potential exists to get hands or fingers injured between the conveyor wheels or rollers and the moving materials or injured between two pieces of materials traveling along the conveyor. In either case, cut, crushed, smashed, and broken hands and fingers can occur. With these types of injuries, guards are seldom the solution for preventing of injuries, but proper use of the conveyors and gloves can help to prevent conveyor-related injuries.

Figure 4-2. Gravity or inertial conveyor (roller type)

GUIDELINES FOR USING MOTORIZED OR POWERED CONVEYORS SAFELY

Since conveyors present a moving hazard to workers, great care is to be taken to insure that operators have access to the on/off switch and that an audible warning exists prior to the conveyor starting.

Workers are prohibited from riding moving chains, conveyors, or similar equipment. Many workers have been caught between the belt and rollers resulting in the loss of fingers, hands, arms, or other severe injuries and even death. (See Figure 4-3.)

Figure 4-3. A typical example of a conveyor related fatality. Courtesy of the Mine Safety and Health Administration

Thus conveyors and other such equipment need to be equipped with emergency cut-off switches, and workers should be protected from the conveyors' moving parts by guards. All nip points and nose or tail pieces should be protected so workers cannot come into contact with moving pulleys, gears, belts, or rollers (see Figures 4-4 through 4-10 for examples of conveyor guarding and safety). Conveyors should be shut down and locked out or tagged out during maintenance and repair work.

The general industry standards do not specifically address conveyors. But be aware that the intent of machine guarding regulations makes it clear that workers should be protected from moving parts and from areas where pinch points or access could result in contact with the movement of any part of the equipment. The construction and mining industries standards do address conveyors' use within the workplace.

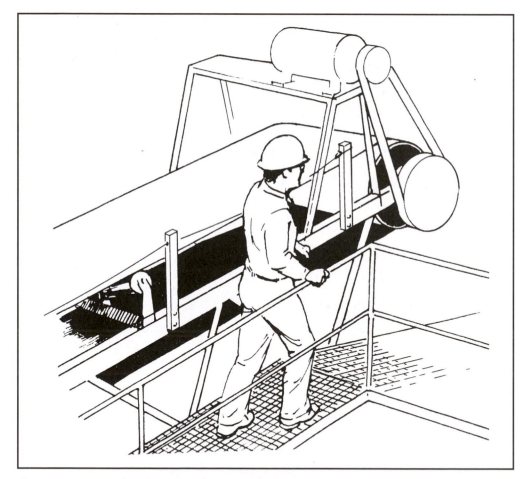

Figure 4-4. Example of conveyor hazards with an unguarded head pulley and drive unit. Courtesy of the Mine Safety and Health Administration

Figure 4-5. Example of an adequately guarded head pulley and drive unit. Courtesy of the Mine Safety and Health Administration

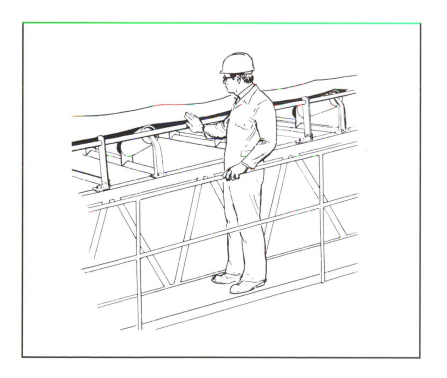

Figure 4-6. A railing is installed along the conveyor and is placed away from the belt to prevent contact. This is adequate guarding. Courtesy of the Mine Safety and Health Administration

Figure 4-7. A fully guarded self-cleaning tail pulley. Courtesy of the Mine Safety and Health Administration

Figure 4-8. A safety stop cord has been installed along the conveyor with a walkway. Courtesy of the Mine Safety and Health Administration

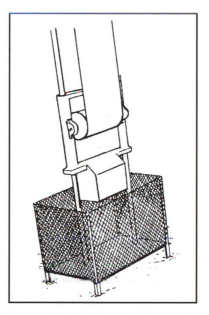

Figure 4-9. Example of a take-up pulley that has been "guarded by location." Courtesy of the Mine Safety and Health Administration

If material can fall from the conveyor, overhead protection needs to be in place. The walkway under the conveyor should have barriers or barricades to prevent individuals from passing through and warning signs should be well placed. For specific questions consult ANSI B20.1-1957.

CONVEYOR REGULATIONS (GENERAL INDUSTRY)

These regulations are for specific types of operations and not intended for application to the totality of the general industry:

Bakeries [(29 CFR 1910.263(d)(7)(I){7}]
Dip Tanks [29 CFR 1910.108(c)(6)]
Electrostatic Spraying [29 CFR 1910.107(h)(7)]
Forging Machines [29 CFR 1910.218(j)(3)]
Sawmills [29 CFR 1910.265 (c)(18)]
Spray Booths [29 CFR 1910.107(b)(7)]
Conveyors [29 CFR 1926.555]

Figure 4-10. An example of a screw conveyor for moving sand.

CONVEYOR SAFETY

Workers should never:

1. Jump across conveyors; whether moving or not.

2. Work on conveyors that are not stopped and locked out and tagged out.

3. Place any part of their body near the moving parts of conveyors.

4. Attempt to ride a conveyor.

5. Shovel material onto a conveyor belt in the opposite direction of the belt's movement.

6. Remove guards or safety devices.

REFERENCES

United States Department of Labor. Mine Safety and Health Administration. *MSHA's Guide to Equipment Guarding for Metal and Nonmetal Mining.* Arlington, VA, 1992.

United States Department of Labor. Occupational Safety and Health Administration. *OSHA 10 and 30 Hour Construction Safety and Health Outreach Training Manual.* Washington, DC, 1991.

United States Department of Labor. Occupational Safety and Health Administration. Construction. *Code of Federal Regulations.* Title 29, Part 1926. Washington, GPO, 1998.

CHAPTER 5

CRANES

Multiple crane worksite

CRANE SAFETY

More than 250,000 crane operators and a very large but undetermined number of other workers and the general public are at risk of serious and often fatal injury due to accidents involving cranes, derricks, hoists, and hoisting accessories. There are approximately 725,000 cranes in operation today in the construction industry, as well as an additional 80,000-100,000 in general and maritime industries.

According to the Bureau of Labor Statistics' Census of Fatal Occupational Injuries, 79 fatal occupational injuries were related to cranes, derricks, hoists, and hoisting accessories in 1993. Occupational Safety and Health Administration's (OSHA) analysis of crane accidents in general industry and construction identified an average of 71 fatalities each year. While we lack adequate worker exposure data to calculate the risk of death for the entire population exposed, the risk of death among crane operators alone is significant, corresponding to a risk of more than one death per thousand workers over a working lifetime of 45 years.

HAZARD DESCRIPTION

In 1992 OSHA reviewed the accident investigation files of 400 crane incidents in general industry and construction over a five year period and identified 354 fatalities, an average of 71 fatalities per year. Although we lack adequate worker exposure data, OSHA found that the risk of death among crane operators is significant and most of these deaths can have commonly identifiable causes. OSHA's analysis also identified the major causes of crane accidents as: boom or crane contact with energized power lines (nearly 45 percent of the cases), under the hook lifting device, overturned cranes, dropped loads, boom collapse, crushing by the counter weight, outrigger use, falls, and rigging failures.

According to the 1987 Bureau of Labor Statistics' (BLS) supplementary data system (23 states reporting), over 1,000 construction injuries were reported to involve cranes and hoisting equipment. However, underreporting of crane-related injuries and fatalities, due to misclassification and a host of other factors, masks the true magnitude of the problem. The 1989 catastrophic tower crane collapse in downtown San Francisco, for example, and the 1993 mobile crane accident near Las Vegas heightened public awareness to the continuing problem of crane accidents. Since crane activities normally occur in urban areas, unsafe equipment and operations present a risk not only to workers, but to the general public as well. Two citizens were killed in San Francisco and three were killed in Nevada. Safe crane operation are both an industry and public concern.

Some cranes are not maintained properly nor inspected regularly to ensure safe operation. Crane operators often do not have the necessary qualifications to operate each piece of equipment safely, and the operator qualifications required in the existing regulations may not provide adequate guidance to employers. Issues of crane inspection and certification and issues of crane operator qualifications and certification need to be further examined.

An agreement was signed by OSHA and the National Commission for the Certification of Crane Operators (NCCCO) in February 1999. For the first time OSHA will accept a training and certification by an outside entity as meeting OSHA's training requirements.

THE REASONS FOR CRANE ACCIDENTS

In our highly mechanized world, cranes are workhorses that have increased productivity in construction, mining, logging, maritime operations, and in the maintenance of production and service facilities. Statistics show, however, that because of inherent hazards which occur during normal working circumstances, a crane can be a very dangerous piece of equipment. Most crippling injuries and deaths from crane accidents can be attributed to 32 hazards that are discussed or summarized in this chapter.

Serious accidents should be examined to identify hazards so that appropriate hazard prevention measures can be initiated. Unfortunately, at the worksite, a crane accident is often labeled as a freak accident because operating personnel are usually unaware of similar repetitive occurrences elsewhere. Employers generally respond to safety requirements when shown that a specific hazard has caused repeated accidents and must be eliminated to avoid future occurrences from the same source.

Many cranes are inadvertently or unintentionally designed with built-in hazards. Although design engineers are very competent, when it comes to crane design, they do not always have sufficient training in the most complex component of crane operation: the people who are going to use the crane. People make errors in judgment when working with cranes that can be used in so many different ways. Too much reliance has been placed upon the ability of those using cranes to make last-minute adjustments and avoid accidents. When safety is included in design, the errors of those who use the equipment can be overcome by the same technology that space engineers use when designing spacecraft.

It is not an insurmountable task to make the workplace safe when using cranes if the crane's limitations and the limitations of those using them are known. Analysis of crane accidents has shown that there was usually no planning of crane use in particular work circumstances. Because the man/machine aspect of design has not received the attention it should have had, cranes arrive at the workplace with inherent hazards. It becomes a challenge to identify these hazards and apply the appropriate OSHA safety measures for their elimination.

HAZARD PREVENTION PLANNING

Preconstruction Planning

Analysis shows that most crane accidents could have been easily prevented had some basic consideration been given to the safe use of cranes and had such considerations been incorporated into the preconstruction planning meeting. The best time to address hazard avoidance is at a preconstruction planning meeting.

Job Hazard Analysis

Before actual craning operations are begun at the job site, a specific job hazard analysis should be made to insure that preconstruction planning is adequate.

Hand Signals

Before any lifts are commenced, all parties, including the crane operator, signalers, riggers, and others involved, must refamiliarize themselves on appropriate hand signals. Often signals vary from job to job and region to region. It is best to assure that everyone is familiar with the hand signals outlined in ANSI/ASME B30.5, Mobile and Locomotive Cranes. OSHA 1926.550(a)(4) state–"Hand signals used by crane and derrick operators are those prescribed by the applicable ANSI standard for the type of crane in use."

Signaling Devices

On lifts where the signalers are outside the direct view of the operator due to elevation or in blind areas, either a telephone or radio is a necessity. Communication needs to be agreed upon prior to lifting.

Lifting Capabilities

During preconstruction planning, any lifting requirements should by analyzed by an engineer competent in this field to establish if the crane to be used has adequate lifting capability.

Rigging Practices

The use of slings to support loads is well-defined in OSHA 1910.184, "Slings." OSHA 1926.251, "Rigging Equipment for Material Handling," defines requirements for rigging equipment.

Controlling the Load

The use of tag lines to control movement of the load is very important. Normally when hoisting a load, the lay or twist in wire rope causes rotation when the load becomes suspended. OSHA 1910.180 (h)(3)(xvi) states: "A tag or restraint line shall be used when rotation of the load is hazardous."

Wire Rope Requirements

It is very important to comply with the crane manufacturer's recommendations for type of wire rope to be used for various hoist lines or pendants.

Annual Inspections

According to OSHA 1910.179(j), 1910.180(d), 1910.181(d), 1917.45(k), and 1926.550(a)(6), all cranes require an annual inspection. There are a number of firms certified to perform these inspections.

Preventive Maintenance

Cranes require continual servicing and preventive maintenance. Programs should be documented consistent with the crane manufacturer's recommendations.

Within the scope of the OSHA, there is usually little opportunity to change the design of existing cranes, but there are measures within OSHA's span of control that can be taken to prevent a hazard from becoming armed and active. In decreasing order of importance, the most effective ways to control hazards are:

- Design to eliminate or minimize the hazard.
- Guard the hazard.
- Give warning.
- Special procedures and training.
- Personal protective equipment.

GENERAL OSHA REQUIREMENTS FOR CRANES AND DERRICKS

Federal regulations covering the use and safety of cranes and derricks are found in several sections of 29 CFR:

- General Industry—Sections 1910.179 through 1910.181.
- Maritime Operations—Sections 1917.45 through 1917.46.
- Longshoring—Section 1918.74.
- Construction—Section 1926.550.

TYPES OF CRANES GENERALLY FOUND IN THE WORKPLACE

- Mobile hydraulic cranes: rough terrain or wheel-mounted telescoping boom. (See Figure 5 -1.)
- Truck-mounted cranes: hydraulic boom or latticework boom. (See Figure 5 -2.)
- Flatbed truck-mounted cranes: hydraulic boom or articulated boom.
- Crawler-mounted latticework boom cranes.
- Overhead track-mounted cranes.
- Monorails and underhung cranes.
- Straddle cranes.
- Fixed cranes: hammerhead tower cranes, pillar cranes, or derricks.

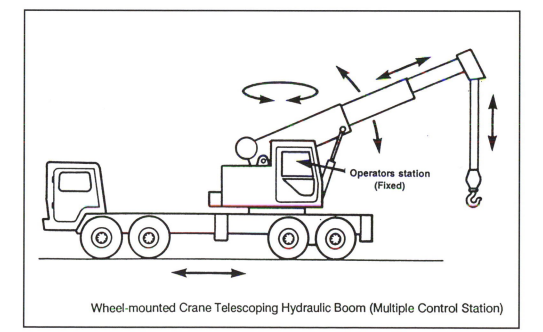

Wheel-mounted Crane Telescoping Hydraulic Boom (Multiple Control Station)

Figure 5-1. Mobile hydraulic crane. Courtesy of the Occupational Safety and Health Administration

Figure 5-2. Truck-mounted crane

ANALYSIS OF HAZARDS COMMON TO MOST CRANES

The breadth of hazards around moveable cranes or derricks is presented in the following section. These are the most common hazards, which cause safety related accidents and incidents around cranes and derricks.

Powerline Contact

Inadvertent contact with a bare, uninsulated high-voltage powerline by any metal part of a crane. Most powerline contacts occur when a crane is moving materials adjacent to or under energized powerlines, and the hoist line or boom touches a powerline. Sometimes the person electrocuted is incidental to the operation or use of the crane: he/she may have been touching the crane or getting on or off of it when the hoist line or boom contacts with an energized powerline.

The key to avoiding powerline contact is prejob safety planning, the greatest accident deterrent available in the workplace. Because of the large number of individuals and employers at a worksite (the landowner, construction management, prime contractor, subcontractors, crane rental firms, electric utilities, etc.), a single individual should have overall supervision, coordination, and authority for the project and project safety. Cranes and powerlines should not occupy the same work area. Contractors should not rely upon electric utility companies to erect either temporary or permanent powerlines at the work site with clearances for cranes. Some work areas with existing powerlines have clearances acceptable for normal roadway traffic but not for cranes. The National Electrical Safety Code clearly set forth that clearances listed are not applicable to construction, maritime activities, mining, etc. where cranes may be used. This is generally a ten-foot diameter area in any direction which should be marked and barricaded.

Human specialists for years have stated that it is beyond the range of normal human performance to:

- Accurately visually judge clearances between a crane and powerlines.

- Observe more than one visual target at a time.

- Overcome the camouflaging characteristics that trees, buildings, and other objects have upon powerlines.

- Cage-type boom guards, insulating links, and proximity warning devices are redundant backup measures and are not substitutes for maintaining a safe clearance.

Overloading

Exceeding the rated capacity or tipping load of a crane while attempting to lift or maneuver a load, which could result in upset or structural failure. Cranes can easily upset from overloading. The margin of safety between the actual tipping load and rated capacity often varies from fifteen to twenty-five percent, based on the type of crane. In some types of cranes, the location of the boom in relation to where the carrier is mounted determines the margin of safety between tipping load and rated capacity.

For a crane mounted on a flatbed truck with the hydraulic boom located directly behind the truck cab, the tipping load over the rear of the flatbed or over the cab can often be as much as twice the rate capacity, but the crane can easily upset when the load is slewed (rotated) to either side.

The mere weight of a boom without a load can create an imbalance and cause some high-reach hydraulic cranes to upset when the boom is positioned at a low angle. This has occurred even with outriggers extended. Today's crane operator is confronted with a number of variables that affect lifting capacity:

- The ability to extend a hydraulic boom, while raising or lowering it, increases the radius very swiftly and reduces lifting capacity quickly.

- Whether the operator chooses to extend or retract outriggers is a variable that affects crane stability.

- The crane's tipping capacity can also vary when the boom is positioned at the various points of the compass or dock in relation to its particular carrier frame.

- In many instances, the operator may not know the actual weight of the load. Often the operator relies upon his perception, instinct, or experience to determine whether the load is too heavy and may not respond fast enough when the crane begins to feel light.

During the last thirty years, solid-state micro-processing electronics for load-measuring systems have evolved. Crane operation is no longer a seat-of-the-pants skill. It requires both planning and training in the use of the latest technology for load-measuring. These load-measuring systems can sense the actual load in relation to the boom angle and length, warn the operator as rated capacity is approached, and stop further movement. Load-measuring systems automatically prevent exceeding the rated capacity at any boom angle, length, or radius. Today most U.S. crane manufacturers are promoting the sale of user-friendly load-measuring systems to avoid this hazard.

Before computerized load-measuring systems, the only control to avoid upset from overload has been an operator's performance using difficult-to-read load charts. Formal training should be provided for all crane operators to assure competency in the use of crane load charts. Providing the operator with both know-how and measuring devices doubles the possibility that load handling equipment will be operated safely.

Failure to use Outriggers/Soft Ground and Structural Failure

Crane upset can occur when an operator does not extend the outriggers, when a crane is positioned on soft ground, and when structural defects exist. Many cranes upset because use of outriggers is voluntary and usually left to the discretion of the operator, who might not perceive a potential hazard such as:

- Sometimes an operator cannot extend the outriggers because of insufficient space or a work circumstance that arises when preplanning is not done.

- Often outrigger pads are too small to support the crane even on hard ground, let alone on soft ground, which poses a problem in itself.

- A mobile hydraulic rough-terrain crane on "rubber" with the outriggers retracted is completely unstable on a side lift.

- Whether a crane is on rubber or is truck-mounted, if outriggers are not extended, the lifting capacity at side angles drops dramatically. As the crane boom rotates on the turntable, upset occurs so quickly that the operator cannot perceive the loss of stability until too late. Sometimes the weight of the boom at a low angle will upset the crane even with the outriggers properly extended.

- In a few instances, outriggers have collapsed because they were not strong enough or had been damaged.

- The rear outrigger on some cranes disengages from the float or pad when the load is first lifted. When the crane cab/boom is swung around 180 degrees, the outrigger can slip from such an unfixed pad connection and cause the crane to upset or the boom to buckle.

Since a high proportion of accidents occur when outriggers are not extended, design changes to overcome this hazard would be desirable. Some aerial basket designs include limit switches to prevent boom movement until outriggers are extended and in place (See Figure 5 - 3.) Outrigger/soil failures occur because either the ground is too soft or the outrigger pads are not broad enough for the soil type. Wet sand can only support 2,000 pounds per square foot, dry hard clay can support 4,000 pounds per square foot, and soil compacted to well-cemented hardpan can support as much as 10,000 pounds per square foot. When poor supporting soil is encountered, or the outriggers have inadequate floats or pads, well-designed blocking or cribbing is needed under the outriggers to extend the base of the outrigger support. ANSI requires that where floats are used on outriggers, they shall be securely attached. It also requires that blocking used to support outriggers shall be strong enough to prevent crushing, be free of defects, and be of sufficient width and length to prevent shifting or toppling under load.

Two-Blocking

The contact of the hoist block or hook assembly with the boom tip causes parting of hoist line and loss of the load, hook, etc., endangering workers below. Both latticework and hydraulic boom cranes are prone to two-blocking. When two-blocking occurs on latticework booms, the hoist line picks up the weight of the boom and lets the pendant guys go slack. Often a great deal of whip is created when walking a crawler crane with a long boom without a load, and the headache ball and empty chokers can drift up to the boom tip. Ordinarily, while busy watching the pathway of travel to avoid any rough ground, which can violently jerk the crane, and cause problems when the crane operator does not watch the boom tip. When a hoist line two-blocks, it assumes the weight of the boom and relieves the pin-up guys of the load. Then,

Figure 5-3. Crane with outriggers extended

if the crane crawler breaks over a rock or bump, the flypole action of a long boom is sufficient to break the hoist line.

The weight of the load plus the weight of the boom on a latticework boom, when combined with a little extra stress from lifting a load, can cause the hoist line to break if two-blocking occurs. The power of hydraulic rams that extend hydraulic booms is often sufficient to break the hoist line if the line two-blocks. In many circumstances, both latticework and hydraulic boom cranes will two-block when the hook is near the tip and the boom is lowered. An operator can forget to release (payout) the load line when extending the boom. When this occurs, the hoist line can be inadvertently broken. If the load line breaks, and it happens to be supporting a worker on a boatswain's chair, several workers on a floating scaffold, or a load above people, a catastrophe can result. Human factors and logic tell us that when an operator must use two controls, one for the hoist and one for the hydraulic boom extension, the chance of error is increased.

Anti-two-blocking devices have been available for years, but industry acceptance has lagged behind when it comes to adopting this device as a preventive measure. There are several ways to prevent two-blocking:

- An electrical sensing device—Attach a weighted ring around the hoist line that is suspended on a chain from a limit switch attached to the boom tip. When the hoist block or headache ball touches the suspended, weighted ring, the limit switch opens and an alarm warns the operator. A device can also be wired to intercede and stop the hoisting. The circuitry is no more complex than an electric doorbell.

- On hydraulic cranes, the hydraulic valving can be sequenced to play out the hoist line, when the boom is being extended, thus avoiding two-blocking.

- When making a lift, sufficient boom length is necessary to accommodate both the boom angle and a full measure of space for rigging, such as slings, spreader bars, and straps. For example, avoid bringing the hook and headache ball into contact with the boom tip, a boom length of 150 percent of the intended lift is required for a boom angle of 45 degrees or more.

- Anti-two-blocking devices should be standard equipment on all cranes. Currently, most new mobile hydraulic cranes are being equipped with anti-two-blocking systems.

Pinchpoint

Pinchpoints are accessible areas where people may be working who can be crushed or squeezed between the carrier frame and the crane cab, or the crane cab and an adjacent wall or other structure. A pinchpoint is created by narrow clearance between the rotating super-structure (cab) of a crane and the stationary carrier frame, or when a crane must be used in a confined space. Another dangerous pinchpoint is the close clearance between the rotation cab/counterweight and a wall, post, or other stationary object. This hazard is inherent in rough terrain, where a crane is truck-mounted, around crawler cranes, and near other mobile cranes. Many people, especially oilers, have been inadvertently crushed in such pinchpoints. Analysis of these occurrences shows that victims were usually "invited" into the danger zone for reasons such as these:

- Access to the water jug.
- Access to the tool box.
- Access to the outrigger controls.
- Access to perform maintenance.
- Access to storage of rigging materials.

In all of the known cases where someone entered the danger zone and was caught in a pinchpoint, the danger zone was outside the crane operator's vision. The survivors stated that they believed the crane operator was not going to rotate or slew the boom at that particular moment.

The swing area of the crane cab and counterweight shall be barricaded against entry into the danger zone. Cranes designed with a 12- to 14-inch clearance between the cab and the carrier frame or crawler tracks may eliminate the hazard and avoid serious crushing injury. Removal of water jugs, tool boxes, and rigging materials from this dangerous area can reduce the incentive to enter the danger zone. Installation of rear view mirrors for the crane operator can provide a redundant safeguard so the operator can see into the turning area of the cab and counterweight. Installation of an audible alarm that will automatically go off when movement begins can alert those working in adjacent areas. Installation of a panic bar or cord that can be activated by anyone entering this blind danger zone can advise the crane operator that someone is in the danger zone so the crane can stop movement until the area is clear.

Moving Parts

Moving parts are all moving or rotating parts within the crane cab, engine compartment, or service area that are accessible to people who must enter or reach into the crane's mechanical housing or compartment. Oilers, maintenance personnel, and crane operators are often injured within the crane cab when caught by one of the many unguarded moving parts. The crane cab is a convenient place to store items essential to crane operations. The cramped machinery space inherent in crane cabs makes it easy for individuals to become entangled with unguarded moving parts.

It is commonly assumed by designers that access doors and panel covers serve as guards to keep people away from moving parts, and that cranes are always shut down when adjustments or servicing is to be done. These assumptions are wrong. Because it is so handy, the cab is commonly used as a convenient storage space for oil, spare parts, tools, lunch boxes, clothing, and other items by maintenance people and others. For this reason, all moving parts that are accessible to anyone need their own guards, inside the cab or a compartment.

Unsafe Hooks

Lifting hooks that do not have latches or have damaged and/or defective latches to safely retain load straps, cables, or chains are often called killer hooks. A safe lifting hook is a critical component when lifting a load. The most unsafe hooks are ones that have no latch to secure the load straps or chain within the throat of the hook. Many hooks have thin sheet metal latches, which are easily damaged or bent, and are totally worthless when it comes to securing the straps safely within the hook. Load hoist blocks are found on some cranes fit flat to a boom. When a boom is in a lowered position, the straps rest deep in the throat of the hook. When the boom is raised in a nearly vertical position, the hook often rotates towards the boom so the opening is positioned downward and the straps can easily slide out of the hook. Another problem inherent with hooks is that, as normal stresses gradually wear away and enlarge the throat, the possibility of the straps slipping from the hook becomes very real. Often a lifting hook is attached to the back of the bucket of a backhoe so slings can be attached and the machine can be used as a lifting device. When the bucket is turned down underneath the boom, the throat of the hook can turn downward, allowing the straps to slip out of the hook. To avoid this hazard, a ring instead of a hook should be used, with a shackle to connect the lifting straps. When two-blocking occurs, the load hook can also rotate up onto the boom tip sheave, allowing the sling to fall from the throat of the hook. When there is no latch on a hook, loose straps without a load can easily be displaced and fall.

Over a hundred patented, positive-type safety latches preventing the hook from opening are available in today's market. Most require the depression of a latch before the hook will open. Use of a vertical swivel allows the hook to remain vertical and prevents dumping the straps from an upended hook. Using wire mousing, or wrapping the hook with wire to hold the load in place of a latch, may secure straps within the throat of a hook. Experience has shown that using wire mousing in place of a hook latch is inadequate, since wire is only good for one use and weakens with each use.

Obstruction of Vision

Safe use of a crane is compromised when the vision of an operator, rigger, or signaler is blocked and they cannot see each other to know what the other is doing. There are two general categories of obstruction of operator's vision. They are obstruction by the crane's own bulk and obstruction by the work environment.

Crane size alone limits the operator's range of vision and creates many blind spots, preventing the rigger, signaler, oiler, and others who are affected by the crane's movement from having direct eye contact with the crane operator. When a cab-controlled mobile crane is moved or travels back and forth, the operator must contend with many blind spots on the right side of the crane. Many situations arise in craning activities, which can almost instantaneously turn what was first considered a simple lift into a life-taking catastrophe:

- Many people are affected by a crane's movement. Welders with their hoods on, carpenters, ironworkers, or those engaged in other crafts may be working in the immediate vicinity of a crane, and may be so preoccupied with their tasks that they become momentarily oblivious of the activity of the crane. They also may be out of the range of vision of the crane operator. This lack of awareness on the part of the crane operator and others contributes to craning accidents.

- In many instances the work environment requires that loads be lifted to or from an area outside of the view of the operator.

- Often a load is lifted several stories high, and the crane operator must rely upon others to assure safe movement of the load being handled.

The key to a safe craning operation is the preplanning of all activities, starting with prejob conferences and continuing with daily planning to address any changes that need to be made. The use of automatic travel alarms is an effective way to give warning to those in the immediate vicinity of crane travel movement. To overcome the hazard of lifting loads from blind spots, the use of radio and telephones is much more effective than relying upon several signalers to relay messages by line of sight.

Sheave Caused Cable Damage

Cable damage is accelerated by the use of sheaves of inadequate diameter or inappropriate groove configurations. This damage usually occurs because the throats or grooves of the sheaves and drums do not meet the minimum diameters recommended by wire rope manufacturers. Wire rope makes today's craning possible. Without it, we would not have our modern cranes. As indispensable as it is, wire rope is very fragile because it is very vulnerable to bending or pinching as it passes over a sheave or pulley. Sheave and drum diameters must be as large as possible. As a wire rope cable passes over a sheave, the outer strands are stretched and compressive forces are exerted on the inner strands. Wire rope is able to stretch in two ways, constructionally and elastically. Constructional stretch is the lay or twist that causes it to rotate. Its elasticity of each separate wire strand that comprises the rope is infinitesimal when compared to that of a rubber band. When a wire strand is stretched beyond the elasticity, it can never return to its former shape and is permanently weakened. Crane cable is often damaged because the angle and width of the throat or groove of the sheave is insufficient, resulting in excessive wear. If a throat is too wide to support the wire rope, the rope can flatten; if it is too narrow, the rope will bind. Excessive wear on crane cables occurs because the throats or grooves of the sheaves and drums found on many cranes do not meet the following minimum diameters recommended by wire rope manufacturers in Table 5-1.

Table 5-1

Minimum Sheave Diameters*

For 6x7 Rope	42 times rope diameter.
For 6x8 Type D	42 times rope diameter.
For 6x19 Rope	30 times rope diameter.
For 6x25 Type B	30 times rope diameter.
For 6x30 Type G	30 times rope diameter.
For 6x37 Rope	18 times rope diameter.
For 8x19 Rope	21 times rope diameter.
For 18x7 Rope	34 times rope diameter.

*Source: the Occupational Safety and Health Administration

Safe throat size is determined by having the diameter of its groove exceed the nominal rope diameter. Table 5-2 gives recommended clearances.

Table 5 -2

Safe Throat Sizes*

Nominal Rope Diameter, Inches	Recommended Groove Diameter Clearance, Inches	
	Minimum Inches	Maximum Inches
1/4 - 5/16	1/64	1/32
3/8 - 3/4	1/32	1/16
13/16 - 1 1/6	3/64	3/32
1 3/16 - 1 1/2	1/16	1/8
1 9/16 - 2 1/4	3/32	3/16
2 5/16 and larger	1/8	1/4

*Source: the Occupational Safety and Health Administration

It must be considered that most cranes do not meet the wire rope manufacturers' design requirements, so excessive wear of crane cables will occur. The actual number of lifts and lowerings must also be considered. Just as a car will run out of gas after a certain number of miles are driven, so will wire rope wear out after a certain amount of travel over the sheave.

Inspect the sheave diameter and the groove to determine if they match the cable size specified by the crane manufacturer. Cable must be inspected daily. Wire rope should be taken out of service when six randomly distributed, broken wires occur in one lay, or three broken wires in one strand in one lay.

A properly lubricated hoist cable will reduce wear.

Cable Kinking

Cable is sometimes damaged during work circumstances, or may be otherwise abused by improper handling, creating kinks and bends. Allowing a wire rope to become slack can cause a crane cable to kink or "birdcage." Slackening of a rope can occur when there is insufficient weight on the load hook, causing the rope to ride up on the boom tip. The slack is then picked up by the heavier weight of the rope extending down from the boom tip to the hoist drum, causing the rope to fall in loops in front of the hoist drum. A kink starts when a loop is formed and is pulled tight: the natural lay of the rope is lost. On cranes with latticework booms, the wire rope can become fouled and kinked as the boom is lowered and the boom hoist assembly is slackened.

Fair leads and sheave guides should be used to prevent a slack line from rolling out of a sheave. Wire rope with kinks or birdcages must be removed from service and destroyed.

Side Pull

The lateral forces imposed upon a boom when the load line is tensioned to either the right or left of the boom can cause side pull. When a crane sits on a slope and picks up or suspends a load, the boom is subjected to side pull. Side pulls sometime arise when two cranes work together to lift a single load.

Cranes must be made level before use. Leveling bubbles and other sensing systems are available. Sensing devices can also be installed to inform the crane operator that a side pull is developing.

Boom Buckling

When a boom is lowered onto or strikes a structure during slewing, the boom cannot sustain side forces, particularly when supporting a suspended load, and will easily collapse. It does not take tremendous force to cause a boom to buckle when it strikes against a structure. Those below are in immediate danger of being killed or injured by the load or the buckling boom. It is better to use tower cranes when the lift has to extend over and into other parts of a building.

Access to Cars, Bridges, and/or Runways

The means by which crane operators, oilers, maintenance personnel, and others have access onto, into, or about the crane to perform their duties is also important. The design of many cranes does not take into consideration the needs of crane operators, oilers, maintenance personnel, and others for safe entry and exit to and from their particular work areas on the crane. Often access is inconvenient, difficult, and unsafe. When those who must work in and about a crane are provided with only marginally safe access due to inadequate ladders, steps, walkways, and platforms, the risk of falls is great. Smooth surfaces over which people must gain access may be coated with oil, grease, or mud, making access slippery. Ladders, steps, walkways, and platforms may also be without sufficient handholds. Attention must be given to safe access.

Nonslip steel treads, surfacing materials, and a wide variety of handholds are available. Therefore, there is little reason to have unsafe access on and off of cranes. Also available are easily assembled and easily placed scaffold planks of adequate capacity, as well as standard railing attachments, and/or safety harnesses and lifelines with safe attachment points to prevent falls.

Control Confusion

Non-uniform control placement or location of controls for the operation of cranes contributes to operator error and inadvertent activation. As with automobiles, there is a wide variety of makes and models of cranes, operated by many different people. On automobiles, brakes, accelerators, and other important controls are always in the same place, so people who rent different cars while traveling do not have a problem with basic controls. With cranes, the location of basic controls often varies, so crane operators, who often change from crane to crane in normal work circumstances, are continually faced to reorient themselves to operate controls in different locations. Additionally, cranes do many things and crane operation is much more complex than automobiles. A crane operator often has to handle multiple controls concurrently.

The absence of uniformity and the many variables of operation create a multitude of opportunities for operator error. A crane operator may be simultaneously slewing (rotating the crane cab), raising the boom, extending the boom, and hoisting a load to a point eight or ten stories high. When the load nears its destination, and just prior to lowering it into place, the operator must stop slewing, raising, extending, and hoisting. Such multiple concurrent actions are a test of operator skill and coordination. Even with the best of operators, there is much room for error.

Since work circumstances often require operators to work with a variety of cranes, controls need to be made uniform and consistent from make to make, model to model, to reduce operator error. When possible, crane operators should be assigned to the same crane, or the same make and model crane, to avoid having to work with dissimilar controls.

Turntable Failure

Turntable failure refers to failure of the rotating mechanism of the crane, which causes it to fall from its turntable. During a heavy lift, the bolts holding the rotating crane hoist structure to the turntable ring or securing the pedestal have been known to fail. An overload usually triggers such failure. Investigations usually show that the connecting bolts or pins had become loose and/or were fatigued from previous overstressing. Crane accident investigations of turntable failure reveal that crane operators are generally unaware of this hazard.

Inspection of the turntable assembly should be part of the annual inspection to determine whether damage or dangerous overstressing has occurred. Also, inspections should be made for a positive program to prevent overloading of the crane.

Removable or Extendible Counterweight Systems

Many cranes utilize removable or adjustable counterweights to make the crane transportable. However, methods of removing or adjusting counterweights are often error-provocative and dangerous. The need for more counterweights increases as cranes are designed to lift greater loads higher and further. Because many of our streets, highways, and bridges have safe load limits, many cranes arrive at the worksite without counterweights attached. The design of some counterweights includes a self-lifting system for loading and unloading the counterweights. Some self-lifting systems are designed only to lift or lower the counterweight from a trailer. If a trailer is unavailable, a second crane is needed to lift the counterweight, if the crane does not have an alternate lifting system. A hydraulic ram is used on a few cranes to increase the capacity of the counterweight. On one model the control panel for the ram is located so that the person operating the controls cannot ascertain, before extension or retraction is begun, whether or not the area surrounding the counterweight is clear. On this particular model, the storage bin for outrigger floats is located under the area of extension or retraction, which creates an invitation for workers to enter this danger zone. In one instance, a worker's head was crushed against the cab by the retracting counterweight. Other means are used on other cranes to extend counterweights further out. Removable or extendible counterweight systems need to be examined to identify the inherent hazards.

ANALYSIS OF HAZARDS COMMON TO SPECIFIC TYPES OF CRANES

Travel Upset in Mobile Hydraulic Cranes (Rough-terrain and Wheel-mounted Telescoping Boom)

Because of a high center of gravity, a mobile hydraulic crane can easily upset and crush the operator between the boom and the ground. This type of crane is easily overturned on road shoulders or other embankments as they are driven from one location to another. When compared to crawler tractors, which can remain stable as much as a 57-degree side slope, mobile hydraulic cranes are rarely stable on side slopes beyond 35 degrees. Because of their versatility with four-wheel drive and four-wheel steer, rough-terrain cranes do encounter slopes of over 35 degrees in their travels that could cause upset. The light-weight, sheet metal cab is also vulnerable to crushing during upset from overloading. The operator has no safe sanctuary in this type of cab to prevent serious injury.

Beginning in the late 1960s, rollover protection system (ROPS) standards were developed by the Society of Automotive Engineers (SAE) for tractors (both crawler and wheel), loaders, graders, compactors, scrapers, water wagons, rear dumps, bottom dumps, fifth-wheel attachments, and various other pieces of equipment. Death and crippling injuries from rollover and falling objects have been substantially reduced because of ROPS. The same technology can be applied to mobile hydraulic cranes so operators can be protected by a crush-resistant cab in the event of upset.

Loss of Control Inadvertent Activation on Two-wheel or Four-wheel Steer Lever on Mobile Hydraulic Cranes (Rough-terrain and Wheel-mounted Telescoping Boom)

Inadvertent shifting from two-wheel steer to four-wheel steering cuts the turning radius in half or causes a crab movement. On some mobile hydraulic, rough terrain cranes, which have the option of two-wheel or four-wheel steer, the lever that engages either two- or four-wheel steer is located dangerously close to the steering wheel (approximately one-half to three-fourths of an inch). An operator, with or without gloves on, can inadvertently brush against the control lever and change the steering from two-wheel to four-wheel steer. Because of the great versatility provided by two-wheel/four-wheel steer, it is easy to drive cranes with this type of steering onto either a soft or sloping shoulder, causing upset.

With only a half to three-quarters of an inch clearance between the steering wheel and the lever that activates two or four-wheel steer, the risk of an operator's inadvertent activation of the two- or four-wheel steer lever is great. In over-the-road travel, such inadvertent change can cause the turning radius to be cut in half in one direction and a crab movement in another direction that would instantaneously make the machine difficult to steer. If the two- or four-wheel drive lever is inadvertently activated by the operator's hand or glove, the operator can lose steering in a critical driving circumstance, even during moderate over-the-road travel on highways at 10 to 15 miles per hour, well below attainable speeds by gear and throttle settings. Inadvertent activation of the two- or four-wheel steer control can be avoided by moving the lever outside the range of the operator's hand or glove so that it cannot be inadvertently struck.

Loss of Stowed Jib Booms on Cantilevered Hydraulic Booms on Mobile Hydraulic Cranes (Rough-terrain and Wheel-mounted Telescoping Boom)

The falling of a stored jib boom from the side of the main boom when the device used for such attachment fall is another hazard. The jib is a useful accessory to lift light loads higher or increase the radius of the lift. When not in use, it is stowed and camed on the side of the main retractable hydraulic boom. For storage, the jib must be totally disconnected from the top extension of the top retractable boom section. It is then rotated 180 degrees so its tip now points down the main boom and is held by a separate support affixed to the main boom.

The boom attachment assembly should be examined to determine whether it is possible for those stowing the jib boom to make errors that would allow it to fall. In many instances a slight modification by the manufacturer can overcome this hazard.

Boom Toppling Over on Lattice-work Boom Cranes

When a lattice-work boom exceeds a 90-degree vertical position, it can easily topple over backwards. (See Figure 5 - 4.) A nearly vertical latticework boom may be toppled by overpulling on the boom hoist cable or by a hoist line that is two-blocked. In some instances, high wind can cause this to occur. Sometimes a crane with its boom in a nearly vertical position will topple over backwards if the crane travels up only a slight slope.

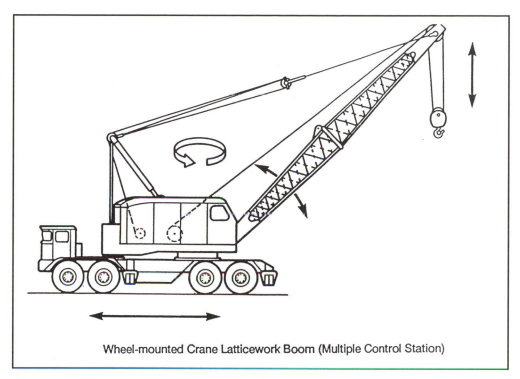

Wheel-mounted Crane Latticework Boom (Multiple Control Station)

Figure 5-4. Lattice-boom crane. Courtesy of the Occupational Safety and Health Administration

Boom stops avoid inadvertent overpulling of either the boom hoist or the two-blocked hoist line. Boom stops do not allow the boom to be raised beyond a safe angle. There are two types of boom stops. One is a mechanical type, which provides opposing or energy-absorbing force, such as the hydraulic boom stops that help avoid the forces created by a sudden release of the load, or by wind. Most mechanical stops are spring-loaded devices and only offer nominal protection. The other type is an electrical interlock sensor-type system that intercedes and halts further boom movement. Many of today's cranes are equipped with both of these mechanical and electrical systems to control this dangerous hazard. During the last few years, because of the almost universal acceptance of both mechanical and electrical boom stops as necessary safety accessories, the toppling over of booms has become less frequent.

Boom Disassembly on Lattice-work Boom Cranes

Improper disassembly can cause a boom to collapse while still suspended, but not blocked, upon those working under the boom removing pins. Lattice-work booms are disassembled for shortening, lengthening, or transporting. Booms collapse occurs on truck or crawler-mounted cranes when the boom is lowered to a horizontal position and suspended from the boom tip with pendant guys, but is not blocked. If the lower pins connecting boom sections are knocked out by workers who are under the boom, the boom can collapse downward upon them, resulting in very serious injuries.

There are at least three circumstances that can lead to accidents when latticework boom sections are being dismantled:

- Workers are unfamiliar with the equipment.

- Poor location is chosen for dismantling.

- Time limit set to complete the function and meet the task deadline is too short.

Planning the boom disassembly location and procedures consistent with the manufacturer's instructions can prohibit hazardous incidents, as can the use of blocking or cribbing on each boom section. Used properly, there are several types of pins that can substantially reduce the risk:

- Double-ended pins that can be removed while standing beside the boom by driving the pin in from the outside.

- Step pins that can only be inserted from inside facing out, and can only be removed by driving from the outside in.

- Welded lugs that prevent pins from being entered the wrong way. This requires the pin to be inserted inside facing out, and can only be removed by driving it from the outside in.

- Screw pins with threads that insert or retract the pin.

- Warnings should be posted at pin connections, and comprehensive written procedures warning of pin hazards and telling how to avoid them should be contained in operators' manuals.

Crane manufacturers used to provide boom sections that were bolted together with bolts placed parallel to the longitudinal dimension of the boom. If the boom were suspended, the boom could not be disassembled as the bolts would bind. Consequently, no boom disassembly collapse injuries were sustained.

Inadvertent Loss of Load on Lattice-work Boom Cranes

Some control systems allow for the unintentional release of the hoist line that supports the load or the pendants that support the boom. Loads are sometimes inadvertently dropped when operator controls are of such design and/or arrangement that a single inadvertent striking of a control lever or pedal can cause an unintentional release of the hoist cable or boom hoist cable. The risk of death or injury is ever present when people are working in the area of boom movement.

Simple design remedies are available that require two distinct motions before the hoist line can be released. The following arrangements avoid inadvertent release:

- One foot movement and a separate hand movement done simultaneously.

- Two distinct hand movements such as:

- Depressing a detent button or hand grasp before the lever can be moved in a forward or backward movement.

- Pressing down on the lever before a forward or backward movement can be initiated.

- Power lowering. But beware, these can become vulnerable if inadvertent release of the control system allows for either power lowering or free fall.

Conductive Cables for Remote Control and Controls Accessible to an Operator Standing on the Ground or on Flatbed Truck-mounted Crane Articulated and Trolley

Hazards occur when an operator is using a remote control, or is standing on the ground using controls mounted on the side of a flatbed truck, is electrocuted when the boom contacts a powerline. (See Figure 5-5.) Remote control boxes are used so the operator can stand on the ground away from the crane and control movement of the boom. Controls are also mounted on the side of the truck bed so the operator can control boom movement from that position. In the event of an inadvertent powerline contact, the remote control box with an electrical cable tether serves as a conductor of power through the operator to the ground. An operator standing on the ground operating the controls on the side of the truck bed is also vulnerable.

Figure 5-5. Flatbed truck-mounted crane

This type of crane, with its remote and side controls, always poses a risk because it is often used at residential building sites to unload materials where the only available space for unloading is under or next to powerlines. Numerous deaths and crippling injuries have been sustained by operators of this type of crane. The use of a radio remote control system can eliminate controls accessible to the operator who is standing on the ground. While not an acceptable primary method of preventing contact with energized conductors, a proximity alarm can be wired to prevent boom operation in locations where powerlines are overhead or nearby.

Unsafe Walkways on Overhead Cranes

Track rails or other walkways are often of unsafe and dangerous design and fail to provide safe access to overhead cranes. How an operator gains access to the crane cab is often ignored in overhead crane design. Often, operators must use the rail track that sits on a beam to get to the cab. Lights to light the lifting area below an overhead crane are sometimes suspended from the underside of the walkway on trap doors that can be opened and raised to replace burned out lights. Operators of overhead cranes are very vulnerable to falls when access on and off the crane is poorly lighted and no special walkways with handrails are provided. Also, when lights need to be replaced, if one of the trap doors is left open and unattended for a short time, the hazard of a fall from an elevation is increased.

Hazards can be avoided by providing safe walkways with handrails for access to the crane cab and maintenance areas. Worker can also avoid falls through open trap doors when manufacturers change the design and eliminate lights suspended from the bottom side of trap doors.

Exposed Electrical Trolleys of Overhead Cranes

On overhead cranes, energized electrical conductors are often exposed and accessible. Contact with energized electrical conductors generally occurs when painting, repairing, or other maintenance is being done, particularly when repair personnel attempt to cross from one overhead crane to another on the same rails or adjacent rails.

The best safeguard is to enclose all exposed and accessible energized electrical conductors. Completely enclosed trolley systems that effectively cover the bare, uninsulated, energized electrical conductors are being marketed.

Absent of Lockout Systems on Overhead Cranes

Lockout systems are necessary on overhead cranes so that they cannot be operated while maintenance or service is being performed. Because overhead crane operators may not always be aware that maintenance or service personnel may be working in close proximity to the legs of the crane as it passes on the raises, a lockout system is needed so maintenance people are not put at risk as they carry out their duties. When repairs are necessary while the crane is in operation, such a lockout system would stop movement of the crane until the maintenance or service work is completed. Many times maintenance or service personnel are working beside a pillar or post that has only a few inches of clearance from the rail on which the crane leg runs. If the operator does not know about the repairs being made or does not see him, the worker can be crushed by the crane leg as it passes along the rail.

Inadvertent movement of the overhead cranes can be controlled by:

• Using temporary bumpers to restrict crane movement in an area where other work is being performed.

• Using a physical electrical lockout device to prevent movement.

Inadequate Stops on Monorails and Underhung Cranes

Monorail systems sometimes have inadequate stops that allowing the monorail's extensions to become disconnected and fall. A monorail crane consists of a fixed track suspended from a ceiling. Suspended from the monorail is often another traveling monorail track from which a traveling chain hoist is suspended. Monorails provide great flexibility to extend the chain hoist either to each side or beyond the end of the fixed monorail. The stops on the fixed monorail or traveling extension are subjected to severe pounding when loads are being moved to the outer limits, and often the traveling rail or chain hoist runs off the end because the stops fail.

Improved rail clamps with well-designed good bumpers and good anchoring systems that will not shear from the normal, constant pounding can prevent this type of failure.

Poor Operator Vision on Straddle Cranes

Straddle cranes are so large that the operator cannot simultaneously view all four legs of the cranes to be sure there are no people or other obstructions in its path. (See Figure 5-6.) The operator station is usually located between the front and rear leg on the left-hand side of the straddle cranes. The dimensions often exceed 30 feet between front and rear wheels and 60 feet between each side. These cranes are often used to handle prestressed concrete beams, ship/truck containers, prefabricated steel components, etc. When lifting a load, the operator has no view from the left side to the right side, when the load being lifted or materials or objects being straddled obstruct his/her view.

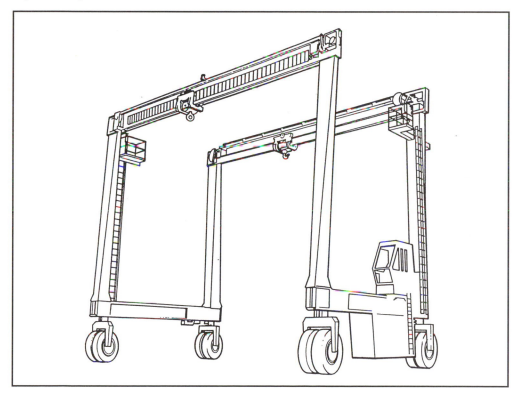

Figure 5-6. Straddle crane. Courtesy of the Occupational Safety and Health Administration

Each leg should be equipped with a travel alarm having a sound level of about 65 decibels at a two-foot distance from the leg to warn workers of impending movement. Additionally, the use of wheel-guard bumpers (cow catchers) as deflectors would push pedestrians harmlessly aside.

Load Loss on Straddle Cranes

A load can be lost due to an imbalance caused by rough movements, rough surfaces, or open lifting hook throats. The use of equalizers or spreader beams to accommodate the long loads usually lifted by straddle cranes often creates an imbalance, especially when lifting bulky or peculiarly shaped objects, and can cause the load to be dropped or fall from one end.

Site preparation is extremely important. A level crane surface must be achieved. Lifting hooks require sturdy latches. A job hazard analysis should be made for each type of function before these cranes are used.

Elevated (Climbing) Tower Cranes

The self-raising or climbing of a tower crane can be its most critical function of operation. When a crane is being raised in a climbing operation, the boom, counterweight, and cab assembly connected to the tower can be inadvertently unbolted without utilizing the temporary latching mechanism for holding the boom and counterweight assembly. The boom and counterweight assembly then becomes precariously balanced on top of the tower, and any movement can cause it to upset and fall. Another failure mode occurs when the high-strength bolts supplied by the crane manufacturer are misplaced and low-strength bolts that can fail under stress are substituted.

Written standard operating procedures for climbing functions would help in preventing this hazard from occurring. The foreman supervising the raising should have a printed checklist, with copies for the crane operator and riggers. This checklist should be followed simultaneously step-by-step by the foreman, the crane operator, and the riggers. Communication either by telephone or radio should be available to verify completion of each step of the printed standard operating procedures before the next step is taken. Connecting bolts and pins must be examined, prior to erection, to determine whether they meet the manufacturer's standards and are not dangerous substitutes.

Footing Failure on Tower Cranes

Placement of a tower crane on a poor foundation can cause it to upset. (See Figure 5-7.) Subsidence of the tower crane footing because of poor soil or other foundation failures renders the crane inoperable as the tower is no longer plumb and will tip. When a tower crane overturns with little warning when its soil or foundation support fails, it can cause serious injury or death to those in the path of its fall. A soils engineer and a structural engineer should be responsible for designing and approving footings for tower cranes.

Wind and Tower or Long Boom Cranes

An unsecured tower crane is very susceptible to wind damage and can fall. Other long boom cranes are also vulnerable to damage from high winds. Wind can easily topple an unsecured tower crane. Craning operations should be discontinued when wind arises.

Weather vaning and wind balance allows, the boom and counterweight to rotate with the wind. In some circumstances the manufacturer may recommend support guys for the tower section. Wind indicators are available to provide warning of dangerous winds. Mobile cranes with long booms should be lowered to the ground when storm warnings predict high winds.

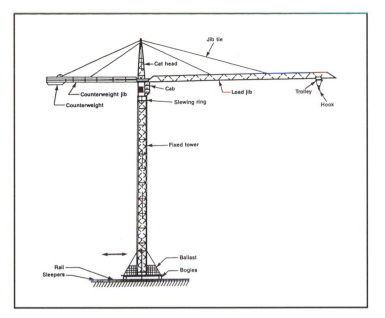

Figure 5-7. Tower crane. Courtesy of the Occupational Safety and Health Administration

COMPETENT PERSONNEL REQUIREMENTS

Operators

As cranes become more sophisticated and are able to lift heavier loads higher further, and faster, more and more electronic systems to monitor every aspect of operator performance are being included in design. The days of total reliance upon seat-of-the-pants operator skills are gone. Human response is not fast enough to cope with many of the rapidly changing circumstances involved in crane operation. Today's crane can be compared to an airplane, not just in terms of cost, but in its complexity of operation, which requires well-qualified professional operators.

Effective training and licensing programs for crane operators offered by some employers include minimum requirements for:

- Education level.
- Apprenticeship training and work experience.
- Classroom crane safety training.
- Thorough knowledge of crane safety references.
- Physical fitness qualifications:
  - Age (mature and intelligent).
  - Vision. ANSI/ASME B30.5, Mobile and Locomotive Cranes, Section 5-3.1.2(b)(1) requires: Vision of at least 20/30 Snellen in one eye and 20/50 in the other, with or without corrective lenses; and Section 5-3.1.2(b)(2) requires: Ability to distinguish colors, regardless of position, if color differentiation is required for operation.

- Hearing. ANSI/ASME.5, Mobile and Locomotive Cranes, Section 5-3.1.2(0)(3) requires: Adequate hearing, with or without hearing aid, for the specific operation.

- Physical stamina.
- Good coordination, reaction, and tested skill level.

- No history of heart problems or other ailments that produce seizures.

- Emotional stability.

- Absence of addictions, etc.

Riggers, Signalers, and Others

Riggers, signalers, and others who work with cranes should have qualifications similar to those of the operator. Just as an unqualified operator can make a life-threatening error during lifting operations, the inappropriate actions of an inexperienced rigger, signaler, or anyone else involved in lifting operations can cause an accident.

CRAWLER LOCOMOTIVE AND TRUCK CRANES (29 CFR 1910.180)

General Requirements

This section applies to crawler cranes, locomotive cranes, wheel-mounted cranes of both truck and self-propelled wheel type, and any variations thereof which retain the same fundamental characteristics. Cranes like these, which are basically powered by internal combustion engines or electric motors, and which utilize drums and ropes, are included with the exception of cranes designed for railway and automobile wreck clearances. Discussion covers requirements of this section are applicable only to machines when used as lifting cranes.

All modern crawler, locomotive, and truck cranes shall meet the design specifications of the American National Standard Institute's Safety Code for Crawler, Locomotive, and Truck Cranes, ANSI B30.5-1968. (See Figure 5-8.)

Only employees selected or assigned by the employer or the employer's representative as being qualified shall be permitted to operate crawler, locomotive, wheel-mounted or truck, and self-propelled type cranes and various related other cranes.

Load Ratings

Where stability governs lifting performance, load ratings have been established for various types of mounting and are given in a table contained in 29 CFR 1910.180. A substantial and durable rating chart with clearly legible letters and figures is to be securely fixed in each crane cabin in a location easily visible to the operator while seated at the control station.

Inspection

Prior to initial use, all new and altered cranes shall be inspected to insure compliance with the provisions of this 29 CFR 1910.180. Inspection procedures for cranes in regular service are divided into two general classifications:

- Frequent inspection—daily to monthly intervals.

- Periodic inspection—one- to 12-month intervals.

Figure 5-8. Crawler crane

Frequent Inspection

All functional operating mechanisms, control systems, safety devices, air and hydraulic systems, cables, chains, rope slings, hooks, and other lifting equipment are to be visually inspected daily for deformation, cracks, excessive wear, twists, stretch, etc. Defective gear shall be replaced or repaired. Running ropes must be inspected monthly and a certification record, including the date of inspection, signature of the person who performed the inspection, and serial number or other identifier must be kept.

Periodic Inspection

Complete inspection of the crane shall be performed at 1-month to 12-month intervals depending on its activity, severity of service, and environmental conditions. The inspection must include the following: deformed, cracked, corroded, worn, or loose members or parts; the brake system; limit indicators (wind, load, etc.); power plant; electrical apparatus; and travel steering, and braking and locking devices.

Inspection Records

Upon formal routine inspection of each crane, certification records including the date of inspection, signature of the person who performed the inspection, and the serial number, or other identifier, of the crane should be made monthly on critical items in use such as brakes, crane hooks, and ropes. This certification record shall be kept on file and readily available to appointed personnel. Daily visual inspections do not require formal records.

Testing

Prior to initial use, all new production cranes shall be tested to insure compliance with the provisions of 29 CFR 1910.180 including the following functions:

- Hoisting and lowering mechanisms.
- Swinging mechanisms.
- Travel mechanisms.
- Safety devices.

Maintenance

After adjustments and repairs have been made, the crane should never be operated until all guards have been reinstalled, safety devices reactivated, and maintenance equipment removed.

Handling the Load

Size of the Load

One of the most significant hazards associated with cranes is overloading. A crane must not be loaded beyond its rated load capacity for any reason except for testing purposes. "Rated load" means the maximum load for which a crane or individual hoist is designed and built by the manufacturer as shown on the equipment name plate.

A common misconception is that a safety factor is built in and that an employer may exceed the rated load. This is not true. A load means the total imposed weight on the load block or hook including the weight of any lifting devices such as magnets, spreader bars, chains and slings.

Attaching the Load

The hoist rope shall not be wrapped around the load. The load shall be attached to the hook by means of slings or other approved devices.

Moving the Load

Some of the requirements for moving loads are stated below.

- The employer must ensure that:
  - The crane is level and, where necessary, blocked properly.
  - The load is well secured and properly balanced in the sling or lifting device before it is lifted more than a few inches.
- Before starting to hoist, the following conditions should be noted:
  - Hoist rope shall not be kinked.
  - Multiple part lines shall not be twisted around each other.

- The hook should be brought over the load in such a manner as to prevent swinging.
- During hoisting, care must be taken that there is no sudden acceleration or deceleration of the moving load and that the load does not contact any obstructions.
- Side loading of booms shall be limited to freely suspended loads. Cranes shall not be used for dragging loads sideways.
- The operator must test the brakes each time a load approaching the rated load is handled by raising it a few inches and applying the brakes.
- Outriggers should be used when the load to be handled at that particular radius exceeds the rated load without outriggers as given by the crane manufacturer.
- Neither the load nor the boom shall be lowered below the point where less than two full wraps of rope remain on their respective drums.
- Before traveling a crane with a load, a designated person must be responsible for determining and controlling safety.
- When rotating the crane, sudden starts and stops should be avoided.
- When a crane is to be operated at a fixed radius, the boom-hoist pawl or other positive locking device must be engaged.

<u>Holding the Load</u>

Operators shall not be permitted to leave their position at the controls while the load is suspended. No person shall be permitted to pass under a load on the hook. If the load must remain suspended for considerable time, the operator shall hold the drum from rotating in the lowering direction by activating the positive controllable means of the operator's station.

Operating Near Electric Power Lines

<u>Clearances</u>

Except where the electrical distribution and transmission lines have been deenergized and visibly grounded at the point of work or when insulating barriers not part of the crane have been erected to prevent physical contact with the lines, the minimum clearance between the lines and any part of the crane or load shall be as shown below.

- For lines rated 50 kV or below, ten feet.
- For lines rated over 50 kV, ten feet plus 0.4 inch for every kV over 50 kV, or twice the length of the line insulator but never less than 10 feet.
- In transit with no load and boom lowered, the clearance shall be a minimum of 4 feet.

<u>Notification</u>

Before commencing of operations near electrical lines, the owners of the lines or their authorized representatives must be notified and provided with all pertinent information.

<u>Overhead Wire</u>

Any overhead wire shall be considered to be an energized line unless and until the owner of the line or the electrical utility authorities indicate that it is not an energized line.

CRANES AND DERRICKS (29 CFR 1926.550)

Rated Loads

The employer must comply with the manufacturer's specifications and limitations applicable to the operation of any and all cranes and derricks. Where manufacturer's specifications are not available, the limitations assigned to the equipment are based on the determinations of a qualified engineer competent in this field, and such determinations will be appropriately documented and recorded. Attachments used with cranes must not exceed the capacity, rating, or scope recommended by the manufacturer.

Rated load capacities, recommended operating speeds, special hazard warnings, and instructions are to be conspicuously posted on all equipment. Instructions or warnings must be visible to the operator while he/she is at the control station. All employees are to be kept clear of suspended loads or loads about to be lifted.

Cranes must deploy outriggers in order to widen the base and be able to lift their intended loads. Warning barriers are erected to keep workers out of the area of operation and the swing radius of the crane itself. (See Figure 5 - 9.)

Figure 5-9. Cranes with outrigger deployed and warning barriers

Hand Signals

Hand signals to crane and derrick operators are those prescribed by the applicable ANSI standard for the type of crane in use. An illustration of the signals must be posted at the jobsite. (See Figure 5-10.)

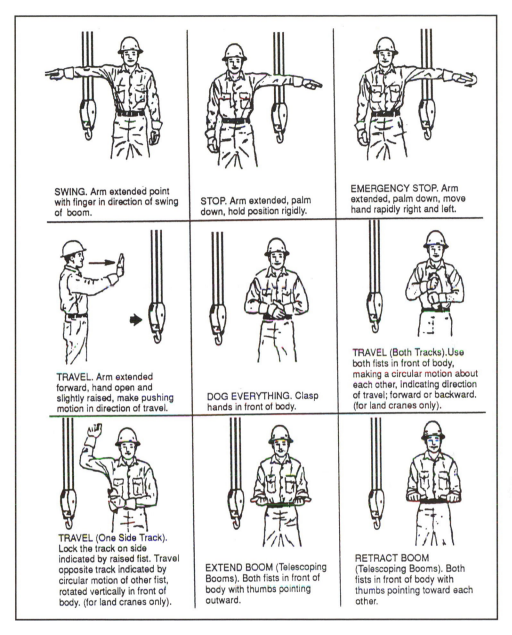

Figure 5-10. Hand signals to be used with cranes and derricks. Permission by the American Society of Mechanical Engineers

Crane Inspections

The employer must designate a competent person to inspect all machinery and equipment prior to each use, and during use, to make sure it is in safe operating condition. (See Figure 5-11 for an example of a crane inspection report form.) Any deficiencies must be repaired, or defective parts replaced, before continued use. A thorough annual inspection of the hoisting machinery must be made by a competent person, or by a government or private agency recognized by the U.S. Department of Labor. The employer is obligated to also maintain a record of the dates and results of inspections for each hoisting machine and piece of equipment.

Figure 5-10. Hand signals to be used with cranes and derricks. Permission by the American Society of Mechanical Engineers (Continued)

MONTHLY CRANE INSPECTION FORM*

Make: _____ Model: _____ Serial # _____

Date of Inspection: _____ Crane Location: _____

Jobsite _____ Inspected by: _____

AREA	N/A	OK	FAULTY	AREA	N/A	OK	FAULTY
General:				**Tracks:**			
Appearance				Chains			
Paint				Sprockets			
Glass				Idlers			
Cab				Pins			
Fire Extinguisher (5BC)				Track Adjustments			
Load Indicator				Roller Path			
Load Charts				Travel Brake			
Hand Signals							
Grease/Oil Leaks							
Boom				**Carrier:**			
Angle Indicator				Condition			
Warning Signs				Brakes			
Steps				Steering			
Guards in Place				Outriggers & Pads			
Hands and Grab Rails				Glass			
Anti-Lock Block				Controls			
Back-up Alarm				Horn			
				Turn Signals			
Engine:				Lights			
Operating Condition				Transmission			
Oil Level and Condition				Frame			
Hour Meter				Battery			
Engine Instruments				License No. (Current)			
Cooling System							
Anti-Freeze				**Boom:**			
Battery Condition				No. or Type			
Hose Condition				Length			
Air System				Swing System			
Pressure				Boom Stops			
Filter				House Rollers			
Compressor				Hook Rollers			
Converters				Swing Gears			
Engine Clutch				Drum Shaft			
Day Tank (Converters)				Clutches			
Hydraulic Reservoir				Main Line			
Oil Pressure				Aux. Line			
Operating Temperature				Brakes			
Transmission Case				Main Line			
				Aux. Line			
Electrical & Control System:				**Boom:**			
Electrical Components & Assemblies				Boom Hoist			
Emergency Stop Switch				Worn Gear			
Overload Switch				Brass Gear			
Master Switches or Drum Controller				Brake			
				Clutches			
				Hydraulic System			

Figure 5-11. Crane Inspection Form

MONTHLY CRANE INSPECTION FORM* (CONTINUED)

AREA	N/A	OK	FAULTY	AREA	N/A	OK	FAULTY
Switches, Contacts & Relays				Pawl			
Main Control Station				Lubrication			
Remote Control Station & Relays				Guards in Place			
Bucket/Ground Controls							
Indicators: Levels: Boom				**Cables & Wire Rope:**			
Angle and Length				Boom Hoist			
Drum Rotation: Load				Load Line			
				Aux./Whip Line			
Boom & Attachments:				Ringer Boom Line			
Number/Type				Load Line Wedge Socket			
Boom Inventory				Kinks & Broken Strands			
Point				(Any Abuse)			
Heel				Jib Pendants			
No. Feet ____							
Load Block (Capacity)				**Records:**			
Hook				Current Certification Posting			
Safety Latch				Operator Instructions			
Boom Pins/Cotter Pins				Preventive Maintenance			
Gantry				Properly Marked Operator			
Equalizer				Controls, Levels &			
Cable Rollers				Diagrams			
Cable Guides				Info. & Warning Decals			
Spreader Bar							
Jib (Type & Length)							
Lubrication							
Jib Inventory				**N/A – Not Applicable**			
Point				**OK – Satisfactory**			
No. Feet ____				**FAULTY****			
Headache Ball (Capacity)				**(**Circles When Repairs Made)**			
Cable Rollers							
Pendant Lines							
Off-Set Links							
Boom Stops							
Auto Boom Stop							
Counterweights							

*Adapted from the Crane Inspection and Certification Bureau, Inc. Monthly Mobile Crane
  Inspection Report

Figure 5–11. Crane Inspection Report Form (Continued)

Wire Rope

Wire rope safety factors are in accordance with American National Standards Institute B 30.5-1968 or SAE J959-1966. Wire rope is taken out of service when any of the following conditions exist:

1. Running ropes, with six randomly distributed broken wires in one lay, or three broken wires in one strand in one lay.

2. There is wear of 1/3 of the original diameter of outside individual wires, or kinking, crushing, bird caging, or any other damage resulting in distortion of the rope structure exists.

3. Evidence exists of any heat damage from any cause.

4. There are reductions from nominal diameter of more than 1/64 inch for diameters up to and including 5/16 inch; 1/32 inch for diameters 3/8 inch up to and including 1/2 inch; 3/64 inch for diameters 9/16 inch up to and including 3/4 inch; 1/16 inch for diameters 7/8 inch to 1 1/8 inch inclusive; 3/32 inch for diameters 1 1/4 to 1 1/2 inch inclusive.

5. In standing ropes, where than two broken wires exist in one lay in sections beyond end connections, or there is more than one broken wire at an end connection.

Guarding

Belts, gears, shafts, pulleys, sprockets, spindles, drums, flywheels, chains, or other reciprocating, rotating, or moving parts or equipment are guarded if such parts are exposed to contact by employees, or otherwise create a hazard. Guarding must meet the requirements of the ANSI B 15.1-1958 Rev., Safety Code for Mechanical Power Transmission Apparatus.

Accessible areas within the swing radius of the rear of the rotating superstructure of the crane, either permanently or temporarily mounted, must be barricaded in such a manner as to prevent an employee from being struck or crushed by the crane.

All exhaust pipes must be guarded or insulated in areas where contact by employees is possible in the performance of normal duties. Whenever internal combustion engine powered equipment exhausts in enclosed spaces, tests shall be made and recorded to see that employees are not exposed to unsafe concentrations of toxic gases or oxygen deficient atmospheres.

All windows in cabs must be made of safety glass, or an equivalent material, which introduces no visible distortion that will interfere with the safe operation of the machine.

Where necessary for rigging or service requirements, a ladder or steps shall be provided to give access to a cab roof. Guardrails, handholds, and steps should be provided on cranes for easy access to the car and cab, conforming to ANSI B30.5. Platforms and walkways are to have anti-skid surfaces.

Fueling

Fuel tank filler pipe must be located in such a position, or protected in such manner, as to not allow spill or overflow to run onto the engine, exhaust, or electrical equipment of any machine being fueled. An accessible fire extinguisher of 5BC rating, or higher, shall be available at all operator stations or cabs of equipment. All fuels must be transported, stored, and handled in a manner, which meets the rules of the construction fire safety regulation. When fuel is transported by vehicles on public highways, Department of Transportation rules contained in 49 CFR Parts 177 and 393 concerning such vehicular transportation are considered applicable.

Electrical Concerns

Except where electrical distribution and transmission lines have been deenergized and visibly grounded at the point of work, or where insulating barriers, not a part of or an attachment to the equipment or machinery, have been erected to prevent physical contact with the lines, equipment, and machines are to be operated proximate to power lines only in accordance with the following:

1. For lines rated 50 kV, or below, minimum clearance between the lines and any part of the crane or load shall be ten feet.

2. For lines rated over 50 kV, minimum clearance between the lines and any part of the crane or load shall be ten feet plus 0.4 inch for each 1 kV over 50 kV, or twice the length of the line insulator, but never less than ten feet.

3. In transit with no load and boom lowered, the equipment clearance is a minimum of four feet for voltages less than 50 kV, and ten feet for voltages over 50 kV, up to and including 345 kV, and 16 feet for voltages up to and including 750 kV.

4. A person is designated to observe clearance of the equipment and give timely warning for all operations where it is difficult for the operator to maintain the desired clearance by visual means.

5. Cage-type boom guards, insulating links, or proximity warning devices may be used on cranes, but the use of such devices does not alter the requirements of any other regulation of this part, even if such device is required by law or regulation.

6. Any overhead wire is considered to be an energized line, unless and until the person owning such line, or the electrical utility authorities, indicate that it is not an energized line and it has been visibly grounded.

7. Prior to work near transmitter towers, where an electrical charge can be induced in the equipment or materials being handled, the transmitter is de-energized or tests are made to determine if electrical charge is induced on the crane. The following precautions are taken, when necessary, to dissipate induced voltage: the equipment is provided with an electrical ground directly to the upper rotating structure supporting the boom, and ground jumper cables are attached to materials being handled by the boom equipment, when an electrical charge is induced while working near energized transmitters. Crews are provided with nonconductive poles having large alligator clips, or other similar protection, to attach the ground cable to the load. Combustible and flammable materials are removed from the immediate area prior to operations.

Modifications

No modifications or additions that affect the capacity or safe operation of the equipment are made by the employer without the manufacturer's written approval. If such modifications or changes are made, the capacity, operation, and maintenance instruction plates, tags, or decals are changed accordingly. In no case is the original safety factor of the equipment to be reduced. The employer shall comply with Power Crane and Shovel Association Mobile Hydraulic Crane Standard No. 2. Sideboom cranes, mounted on wheel or crawler tractors, must meet the requirements of SAE J743a-1964.

Crawler, Locomotive, and Truck Cranes

Crawler, locomotive, and truck cranes must have positive stops on all jibs to prevent their movement of more than five degrees above the straight line of the jib and boom on conventional type crane booms. The use of cable type belly slings does not constitute compliance with this rule. All crawler, truck, or locomotive cranes in use must meet the applicable requirements for design, inspection, construction, testing, maintenance, and operation as prescribed in the ANSI B30.5-1968, Safety Code for Crawler, Locomotive and Truck Cranes. However, the written, dated, and signed inspection reports and records of the monthly inspection of critical items prescribed in section 5-2.1.5 of the ANSI B30.5-1968 standard are not required. Instead, the employer shall prepare a certification record, which includes the date the crane items were inspected, the signature of the person who inspected the crane items, and a serial number, or other identifier, for the crane inspected. The most recent certification record shall be maintained on file until a new one is prepared.

Hammerhead Tower Cranes

For hammerhead tower cranes, adequate clearance is to be maintained between moving and rotating structures of the crane and fixed objects, in order to allow for safe passage of employees. Each employee required to perform duties on the horizontal boom of the hammerhead tower cranes is protected against falling by guardrails, or by a personal fall arrest system. Buffers are provided at both ends of travel of the trolley. Cranes mounted on rail tracks are equipped with limit switches limiting the travel of the crane on the track, and stops or buffers at each end of the tracks. All hammerhead tower cranes in use are required to meet the applicable requirements for design, construction, installation, testing, maintenance, inspection, and operation as prescribed by the manufacturer.

Overhead and Gantry Cranes

For overhead and gantry cranes, the rated load of the crane is plainly marked on each side of the crane, and if the crane has more than one hoisting unit, each hoist must have its rated load marked on it or its load block; this marking must be clearly legible from the ground or floor. Bridge tracks are equipped with sweeps which extend below the top of the rail and project in front of the truck wheels. Except for floor-operated cranes, a gong or other effective audible warning signal shall be provided for each crane equipped with a power traveling mechanism. All overhead and gantry cranes in use are to meet the applicable requirements for design, construction, installation, testing, maintenance, inspection, and operation as prescribed in the ANSI B30.2.0-1967, Safety Code for Overhead and Gantry Cranes.

Derricks

All derricks in use shall meet the applicable requirements for design, construction, installation, inspection, testing, maintenance, and operation as prescribed in American National Standards Institute B30.6-1969, Safety Code for Derricks. For floating cranes and derricks, when a mobile crane is mounted on a barge, the rated load of the crane must not exceed the original capacity specified by the manufacturer. A load-rating chart, with clearly legible letters and figures, must be provided with each crane, and securely fixed at a location easily visible to the operator. When load ratings are reduced to stay within the limits listed for a barge with a crane mounted on it, a new load-rating chart shall be provided.

Floating Cranes and Derricks

Mobile cranes on barges must be positively secured. For permanently mounted floating cranes and derricks, when cranes and derricks are permanently installed on a barge, the capacity and limitations of use are based on competent design criteria. A load rating chart, with clearly legible letters and figures, must be provided and securely fixed at a location easily visible to the operator. Floating cranes and floating derricks in use must meet the applicable requirements for design, construction, installation, testing, maintenance, and operation as pre-scribed by the manufacturer. The employer shall comply with the applicable requirements for protection of employees working onboard marine vessels.

OVERHEAD AND GANTRY CRANES (29 CFR 1910.179)

General Requirements

Overhead and gantry cranes, including semi-gantry, cantilever gantry, wall cranes, storage bridge cranes, and others having the same fundamental characteristics, are covered in this section. (See Figure 5 -12.) Overhead and/or gantry cranes may not be modified and re-rated unless the modifications and the supporting structure are checked thoroughly for the new rated load by a qualified engineer or the equipment manufacturer. It is not unusual to find instances of overhead or gantry cranes where it is claimed that the lifting capacity is increased simply by installing a new rated load sign on the bridge of the crane.

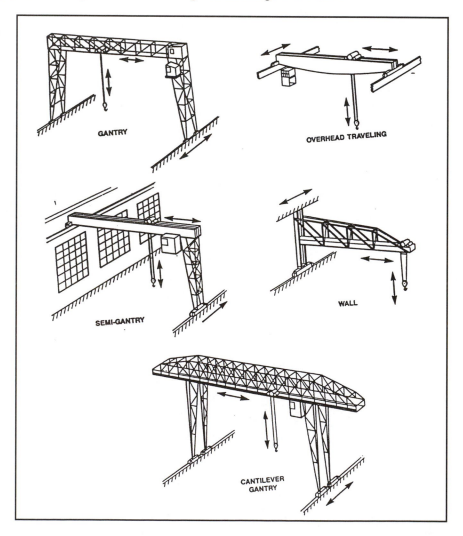

Figure 5-12. Example of overhead and gantry crane. Courtesy of the Occupational Safety and Health Administration

The rated load of the crane is to be plainly marked on each side of the crane. If the crane has more than one hoisting unit, each hoist must have its rated load marked on it or on its load block. The potential for overloading the crane increases if the hook-up person and/or the operator does not know the rated capacity. Only employees selected or assigned by the employer or the employer's representative as being qualified to operate a crane shall be permitted to do so.

Cabs

A cab-operated crane is an overhead or gantry crane controlled by an operator in a cab located on the bridge or trolley. The general arrangement of the cab and the location of control and protective equipment should be such that all operating handles are within convenient reach of the operator when facing the area to be served by the load hook, or while facing the direction of travel of the cab. The arrangement must allow a full view of the load hook in all positions.

The access to all cab-operated cranes must be inspected thoroughly. Serious injuries have occurred because of the following three conditions:

- There was no conveniently placed fixed ladder, stairs, or platform provided to reach the cab or bridge footwalk. It is unacceptable and poses a significant hazard to allow employees to board a crane via climbing over guardrails, over, under, and around building structures, energized hot rails, portable ladders or movable platforms.

- There was a gap exceeding 12 inches between a fixed ladder, stairs, or platform, and access to the cab or bridge footwalk.

- The fixed ladder used as access to the crane did not meet the American National Standard Safety Code for Fixed Ladders, ANSI A14.3-1956. With access ladders, there were no cages provided for ladders over 20 feet in unbroken length, offset platforms were not provided, or the ladders themselves are not maintained in a safe condition.

Footwalks and Ladders

Where sufficient headroom is available on cab-operated cranes, a footwalk shall be provided on the drive side along the entire length of the bridge of all cranes having the trolley running on the top of the girders.

Significant hazards exist for maintenance and inspection personnel if no footwalk is provided. For example, work may be performed from portable ladders, off the main bridge girder itself without protection against falling to the floor below, or from the trolley platform itself with the same potential for falling to the floor below. Maintenance managers and supervisors shall check very thoroughly the maintenance procedures followed in those cases where cab-operated cranes are not provided with bridge footwalks. This, of course, also applies to all other types of cranes where no footwalk is provided and in those cases where bridge footwalks cannot be provided because sufficient headroom is not available.

Bridge footwalks, where provided, shall be of rigid construction and designed to sustain a distributed load of at least 50 pounds per square foot. In many older workplaces, serious hazards are associated with the bridge footwalk itself. This area cannot be inspected from the floor, and safety people must climb onto the crane to properly document hazardous conditions. It is quite common to find bridge footwalks not continuous or permanently secured. Maintenance employees, inspection personnel, and the crane operator must go on the bridge footwalk at various times. It must be remembered that even though at the time of inspection, no employees may be on the bridge footwalks of cranes. A common and serious hazard exists where standard railings have not been provided on all open sides of the bridge footwalk. In addition, toeboards must be installed. The standard railing provisions apply to all sides of the bridge footwalk including the inside edge next to the bridge girders if a fall potential exists.

All gantry cranes are to be provided with a ladder or a stairway that extends from the ground to the footwalk or the cab platform. It is not permitted to board a gantry crane via portable ladders, structure of the crane, end, or other methods.

Any ladder provided on an overhead or gantry crane is to be permanently and securely fastened in place and also shall be in compliance with 29 CFR 1910.27 of the OSHA standards. Damaged, loose, improperly maintained, or unguarded fixed ladders are common.

Stops, Bumpers, Rail Sweeps and Guards

Stops

A "stop" is a device to limit travel of a trolley. This device normally is attached to a fixed structure and normally does not have energy-absorbing ability. Every overhead or gantry crane, where the trolley runs on top of the bridge girder, shall be provided with stops at either end of the limits of the travel of the trolley. These stops shall be fastened to resist forces applied when contacted, and if the stop engages the tread of the wheel of the trolley, it shall be of a height at least equal to the radius of the wheel.

One obvious hazard, related to improperly applied trolley stops or no trolley stops at all, is that the trolley could be run off the trolley runway. Other hazards associated with this condition are numerous and present serious injury potential to the employees on the floor below. The trolley itself could fall to the floor, parts of the trolley could come off the crane structure and hit employees below, the load itself could be dropped, or at a minimum, cause unexpected movement of the load, and finally, if the trolley contacted the bridge runway conductors, the entire crane could itself be energized. The only practical method to inspect for this condition is the boarding of the overhead or gantry crane and walking out on the bridge footwalk to look and see if the trolley stops are there and installed properly.

Modern cranes must meet or exceed the design specifications of the ANSI Safety Code for Overhead and Gantry Cranes, ANSI B30.2.0-1967. A similar hazardous condition is the failure to re-install crane runway stops at the ends of the limits of travel of the runway. Conditions which could and have occurred in many overhead crane installations include the malfunctioning of controllers that become stuck in the open position, and some cranes run off the ends of the bridge runway, often through building walls.

Bridge and Trolley Bumpers

A "bumper" (buffer) is an energy-absorbing device for reducing impact when a moving crane or trolley reaches the end of its permitted travel, or when two moving cranes or trolleys come in contact. Overhead or gantry crane bridges shall be provided with bumpers unless the crane travels at a slow rate of speed and has a faster deceleration rate due to the use of sleeve bearings, or is not operated near the end of bridge travel, or is restricted to a limited distance by the nature of the crane operation, and there is no hazard of striking any object in this limited distance.

A common condition, which will be observed on many overhead or gantry cranes, is that bumpers were not provided where required. However, many times bumpers are provided that do not have sufficient energy-absorbing capacity to stop the crane when traveling at a speed of at least 40 percent of the rated load speed, or the bumpers are not designed and installed as to minimize parts falling from the crane in case of breakage. The hazards of not providing bridge bumpers with energy-absorbing capacities are that when an overhead or gantry crane contacts another crane on the same runway or contacts the bridge stops at the ends of the runway, a shock load is transmitted to the lifting mechanisms, which could cause a potential dropping of the load. In addition, the constant striking of a crane against another object without energy-absorption buffers causes weakening of the bridge and end structures, which could eventually cause cracks in the webbing and lead to failure of the crane structure.

Trolleys shall also be provided with bumpers unless the trolley travels at a slow rate of speed, it is not operated near the ends of the trolley travel, or it is restricted to a limited distance on the trolley runway and there is no hazard of striking any object in this limited distance. If there is more than one trolley operated on the same trolley runway, each trolley shall be provided with bumpers on its adjacent ends. If bumpers are installed on the trolley, they shall be designed and installed to minimize parts falling from the trolley in case of breakage, and shall be energy-absorbing. It must be emphasized that both trolley stops and bumpers shall be provided where required.

Rail Sweeps

Bridge end truck wheels shall be provided with sweeps that extend below the top of the rail and project in front of the truck wheels. Their lack is a very common condition. Also note that this requirement does not apply to the trolley end truck wheels.

By not having rail sweeps, for example, maintenance equipment can be left on the bridge runway rails and, as the crane travels into this area, could be derailed, causing an unintended movement of the load, a shock load, and potential dropping of the load. This also applies to gantry and semi-gantry cranes, where the truck wheels run on a rail usually located on the floor or working surface.

Guards

Hoisting ropes on overhead and gantry cranes must be inspected closely to make sure they do not run near other parts where they could make fouling or chafing possible. If they do run near, guards shall be installed to prevent this condition.

The hazard here is obvious. If a hoisting rope is chafing over a long period of time, it will eventually wear through, or break, and drop the load to the floor below. Also, guards must be provided to prevent contact between bridge conductors and hoisting ropes if they could come into contact. Bridge conductors are almost always located on the inside flange of bridge girders and provide power to the trolley.

With overhead or gantry cranes, it is very common to find exposed moving parts not properly guarded. Examples of this are gears on or near the bridge footwalk, shaft ends on bridge motors usually located on the bridge footwalks, and chain and chain sprockets.

There have been several reported fatalities associated with maintenance employees working on bridge footwalks who were drawn into open gears, projecting shaft ends, and chain and sprocket drives.

Brakes

Each independent hoisting unit of every crane shall be provided with a holding brake, applied directly to the motor shaft or some part of the gear grain. On cab-operated cranes with the cab on the bridge, a bridge brake shall be provided. On occasion, you will find operators of cab-operated overhead cranes that are not equipped with bridge brakes will stop the motion of the bridge by plugging. Plugging is reversing the direction of the bridge motor through the controller. This is a very hazardous practice, especially if it is the only method of stopping a crane. If power to the crane is lost for any reason, the crane will be unstoppable.

All floor, remote and pulpit-operated crane bridge drives shall be provided with a brake or non-coasting mechanical drive, i.e., the crane must be able to stop quickly. It should be noted that overhead or gantry cranes with a cab on the trolley should also be provided with a trolley brake. However, under most conditions trolleys are not be required to have a brake.

Electrical Equipment

All wiring and equipment on overhead or gantry cranes shall comply with the applicable electrical sections of Subpart S. On floor-operated cranes where a multiple conductor cable is used with a suspended push-button station, the station shall be supported in some manner that will protect the electrical conductor against strain. This condition can be abated simply by installing a chair or cable from an upper support to the push-button station to take the strain off the conductor.

Pendant control boxes also must be clearly marked for identification of functions. Lack of clear labeling is quite common. The hazard is that inexperienced operators or supervisory personnel operating the pendant crane may not know the various functions of the push-button station and cause an unexpected movement of the crane. Only designated personnel are permitted to operate a crane per 1910.179(b)(8).

One of the most serious hazards associated with cranes arise from the lack of compliance with the provision requiring that the hoisting motion of all electric traveling cranes be provided with an over-travel limit switch in the hoisting direction. An over-travel "limit switch" or simply a "limit switch" is a switch that disconnects the power to the drive motor and stops the load if that load is raised above a certain point. Many fatalities and serious injuries have occurred when cranes were not provided with a limit switch or the limit switch malfunctioned.

If a limit switch is not provided, or if there is a malfunctioning limit switch, the hoist block could run up into the lifting beam or rope drum, severing the cable and dropping the entire assembly and any load on the hook to the floor below. Therefore, all inspections of an overhead crane must include a close examination of the limit switch.

Hoisting Equipment (Sheaves and Hoisting Ropes)

Sheaves are grooved pulleys that carry hoisting ropes on overhead cranes. Sheave grooves must be smooth and free from surface defects. Sheaves in the bottom blocks shall be equipped with close-fitting guards to prevent ropes from becoming fouled when the block is lying on the ground with the rope loose. This common condition can be readily observed while watching the crane in operation. When guards are not installed, the hoisting rope can come off the sheave groove and become entangled on the shaft, creating binding or shaving of the hoisting rope. Rope fouling can also occur when there is a slack cable, such as when a load block rests on top of a load.

Regarding the hoisting rope itself, there must be at least two wraps remaining on the hoist drum when the hook is in its extreme low position.

Inspection

Prior to initial use, all new and altered cranes shall be inspected to insure compliance with the provisions of this section. Inspection procedures for cranes in regular service are divided into two general classifications:

- Frequent inspection – daily to monthly intervals.
- Periodic inspection – 1- to 12-month intervals.

Frequent Inspection

All functional operating mechanisms, air and hydraulic systems, chains, rope slings, hooks, and other lifting equipment shall be visually inspected daily. Chains, cables, ropes, hooks, etc., on overhead and gantry cranes shall be visually inspected daily for deformation, cracks, excessive wear, twists, stretch, etc., and defective gear shall be replaced or repaired.

Hooks and chains shall be visually inspected daily, although monthly records must be kept, which includes the date of inspection, the signature of the person who performed the inspection, and the serial number or other identifier for each part. Running ropes must be inspected monthly with a certification records documenting the date of inspection, the signature of the person who performed the inspection, and the serial number or other identifier.

Periodic Inspection

Complete inspection of the crane shall be performed at 1-month to 12-month intervals depending on its activity, severity of service, and environmental conditions where it is used. The inspection shall document any deformed, cracked, corroded, worn, or loose members or parts. It should also cover the brake system, limit indicators (wind, load, etc.), power plant, and electrical apparatus.

Testing

Prior to initial use, all new and altered cranes shall be tested to assure compliance with safety requirements covering the following functions:

- Hoisting and lowering.
- Trolley travel.
- Bridge travel.
- Limit switches, locking, and safety devices.

The trip setting of hoist limit switches shall be determined by tests with an empty hook traveling in increasing speeds up to the maximum speed. The actuating mechanism of the limit switch shall be located so that it will trip the switch, under all conditions, in sufficient time to prevent contact of the hook or hook block with any part of the trolley.

Maintenance

A preventive maintenance program based on the crane manufacturer's recommendations shall be established.

Handling the Load

A serious hazard associated with cranes is overloading. A crane shall not be loaded beyond its rated load capacity for any reason except for test purposes. "Rated load" means the maximum load for which a crane or individual hoist is designed and built by the manufacturer as shown on the equipment name plate.

A common and dangerous misconception is that a safety factor is built into the load capacity rating and that an employee may exceed the manufacturer's rated load up to this safety factor. This is not true. A load means the total superimposed weight on the load block or hook and shall include any lifting devices such as magnets, spreader bars, chains, and slings.

Every load lifted by a crane shall be well secured and properly balanced in the sling or lifting device before it is lifted more than a few inches. In some cases, the operator will find the load is balanced but not secured.

To prevent swinging of a load, the hook must be brought directly over the load when the attachment is made. In addition, no employee is permitted on the load or hook or lifting device while hoisting, lowering, or traveling.

The operator of a crane shall avoid carrying loads over other personnel. This hazard is increased significantly when using a magnet or a vacuum device to lift scrap material.

Finally, operators of cranes are not permitted to leave their positions at the controls while the load is suspended. This again includes suspended lifting devices such as magnets or vacuum lifts. At the beginning of each operator's shift, the upper limit switch of each hoist shall be tried out under no-load conditions. Additionally, attached to it shall be a grounded three-prong type permanent receptacle, not exceeding 300 volts.

Overhead and Gantry Cranes (29 CFR 1926.550)

To summarize regulatory policy for construction sites using overhead or gantry cranes, the rated load of the crane must be plainly marked on each side of the crane. If the crane has more than one hoisting unit, each hoist has its rated load marked on it or its load block. This marking must be clearly legible from the ground or floor. Bridge trucks are equipped with sweeps, which extend below the top of the rail and project in front of the truck wheels. Except for floor-operated cranes, a gong or other effective audible warning signal shall be provided for each crane equipped with a power traveling mechanism. All overhead and gantry cranes in use must meet the applicable requirements for design, construction, installation, testing, maintenance, inspection, and operation as prescribed in the ANSI B30.2.0-1967, Safety Code for Overhead and Gantry Cranes.

REFERENCES

Moran, Mark McGuire. *Construction Safety Handbook.* Government Institutes Inc. Rockville, MD, 1996.

Reese, C. D. and J.V. Eidson. *Handbook of OSHA Construction Safety & Health.* Boca Raton, FL: CRC/Lewis Publishers, 1999.

United States Department of Energy. *Hoisting and Rigging Manual.* Washington, DC, 1991.

United States Department of Labor. Occupational Safety and Health Administration. *OSHA 10 and 30 Hour Construction Safety and Health Outreach Training Manual.* Washington, DC, 1991.

United States Department of Labor. Occupational Safety and Health Administration. General Industry. *Code of Federal Regulations.* Title 29, Part 1910. Washington, GPO, 1998.

United States Department of Labor. Occupational Safety and Health Administration. Construction. *Code of Federal Regulations.* Title 29, Part 1926. Washington, GPO, 1998.

United States Department of Labor. Occupational Safety and Health Administration. *OSHA Technical Manual.* Chapter 13 – Cranes and Derricks. Washington, DC, 1990

United States Department of Labor. Occupational Safety and Health Administration. Office of Training and Education. *OSHA Voluntary Compliance Outreach Program: Instructors Reference Manual.* Des Plaines, IL, 1993.

CHAPTER 6

DERRICKS

Derrick on a tow truck

A "derrick" is an apparatus consisting of a mast, or equivalent member, held at the head by guys or braces, with or without a boom, for use with a hoisting mechanism and operating ropes. The Occupational Safety and Health Administration (OSHA) regulation that applies to derricks is found in 29 CFR 1910.181. The general requirements for derricks apply to guy, stiffleg, basket, breast, gin pole, Chicago boom, and A-frame derricks of the stationary type. Each is capable of handling loads at variable reaches and powered by hoists through

93

systems of rope reeving. Each is used to perform lifting hook work, single or multiple line bucket work, grab, grapple, and magnet work. (See Figure 6-1.) Derricks may be permanently installed for temporary use as in construction work. This regulation also applies to any modification of these types of derricks that retain their fundamental features, except floating derricks. All modern derricks shall meet the design specifications of the American National Standard Safety Code for Derricks, ANSI B30.6-1969. Only employees selected by the employer or employees as qualified shall be permitted to operate derricks.

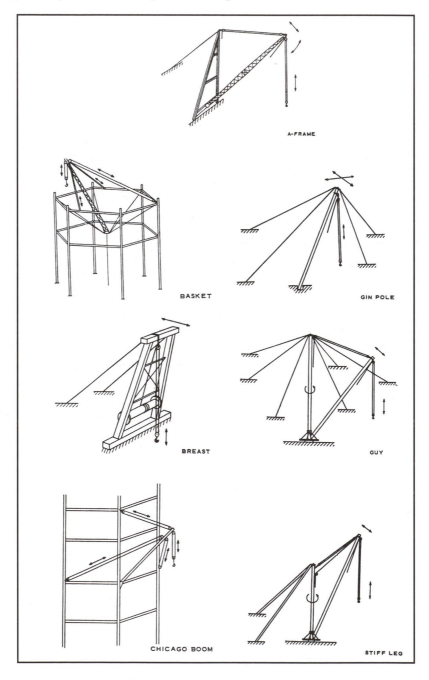

Figure 6-1. Examples of types of derrick. Courtesy of the Occupational Safety and Health Administration

DESCRIPTION OF THE COMMON TYPES OF DERRICKS

"A-frame derrick" means a derrick in which the boom is hinged from a cross member between the bottom ends of two upright members spread apart at the lower ends and joined at the top; the boom point secured to the junction of the side members, and the side members are braced or guyed from this junction point.

A "basket derrick" is a derrick without a boom similar to a gin pole, with its base supported by ropes attached to corner posts or other parts of the structure. The base is at a lower elevation than its supports. The location of the base of a basket derrick can be changed by varying the length of the rope supports. The top of the pole is secured with multiple reeved guys to position the top of the pole to the desired location by varying the length of the upper guy lines. The load is raised and lowered by ropes through a sheave or block secured to the top of the pole.

"Breast derrick" means a derrick without boom. The mast consists of two side members spread farther apart at the base than at the top and tied together at top and bottom by rigid members. The mast is prevented from tipping forward by guys connected to its top. The load is raised and lowered by ropes through a sheave or block secured to the top crosspiece.

"Chicago boom derrick" means a boom which is attached to a structure, an outside upright member of the structure serving as the mast, and the boom being stepped in a fixed socket clamped to the upright. The derrick is complete with load, boom, and boom point swing line falls.

A "gin pole derrick" is a derrick without a boom. Its guys are so arranged from its top as to permit leaning the mast in any direction. The load is raised and lowered by ropes reeved through sheaves or blocks at the top of the mast.

"Guy derrick" means a fixed derrick consisting of a mast capable of being rotated, supported in a vertical position by guys, and a boom whose bottom end is hinged or pivoted to move in a vertical plane with a reeved rope between the head of the mast and the boom point for raising and lowering the boom, and a reeved rope from the boom point for raising and lowering the load.

"Shearleg derrick" means a derrick without a boom and similar to a breast derrick. The mast, wide at the bottom and narrow at the top, is hinged at the bottom and has its top secured by a multiple reeved guy to permit handling loads at various radii by means of load tackle suspended from the mast top.

A "stiffleg derrick" is a derrick similar to a guy derrick except that the mast is supported or held in place by two or more stiff members, called stifflegs, which are capable of resisting either tensile or compressive forces. Sills are generally provided to connect the lower ends of the stifflegs to the foot of the mast.

LOAD RATINGS

Permanent Installations

For permanently installed derricks with fixed lengths of boom, guy, and mast, a durable and clearly legible rating chart shall be securely affixed where it is visible to the operator. The chart shall include:

- Manufacturer's approved load ratings at corresponding ranges of boom angle or operating radii.

- Specific lengths of components on which the load ratings are based.

- Required parts for hoist reeving.

Non-Permanent Installations

For nonpermanent installations, the manufacturer shall provide sufficient information from which capacity charts can be prepared for the particular application. The capacity charts must be located at the derricks or the jobsite office.

INSPECTIONS

Prior to initial use, all new and altered derricks shall be inspected to insure compliance with the provisions of 29 CFR 1910.181. Inspection procedures for derricks in regular service is divided into two general classifications:

1. Frequent inspection—daily to monthly intervals.

2. Periodic inspection—1- to 12-month intervals.

Frequent Inspection

All functional operating systems, control systems, safety devices, chords and lacings, tension in guys, plumb of the mast, air and hydraulic systems, rope reeving, hooks, and electrical apparatus must be visually inspected daily. Running ropes shall be inspected monthly and a certification record shall be kept on file where readily available.

Periodic Inspection

Complete inspection of the derrick shall be performed at 1-month to 12-month intervals, depending on its activity, severity of service, and environmental conditions. The inspection report must include the following: deformed, cracked, corroded, worn or loose members or parts, power plant, and foundation or supports.

TESTING

Prior to initial use, all new and altered derricks must be tested to insure compliance with 29 CFR 1910.181 including:

- Load hoisting and lowering.

- Boom up and down.

- Swing.

- Operation of clutches and brakes of hoist.

All anchorages shall be approved by the appointed person doing the inspection. Rock and hairpin anchorages may require special testing.

MAINTENANCE

A preventive maintenance program, based on the derrick manufacturer's recommendations, shall be established.

HANDLING THE LOAD

Size of the Load

No derrick shall be loaded beyond the manufacturer's rated load limit.

Attaching the Load

The hoist rope may not be wrapped around the load. The load must be attached to the hook by means of slings or other suitable devices.

Moving the Load

Here are some requirements for moving loads:

- The load shall be well secured and properly balanced in the sling or lifting device before it is lifted more than a few inches.
- Before starting to hoist, the following conditions must be noted:
  - Hoist rope shall not be kinked.
  - Multiple part lines may not be twisted around each other.
  - The hook shall be brought over the load in such a manner as to prevent swinging.
- During hoisting, care must be taken that:
  - There is no sudden acceleration or deceleration of the moving load.
  - The load does not contact any obstructions.
- The operator shall test the brakes each time a load approaching the rated load limits is handled by raising it a few inches and applying the brakes.
- Neither the load nor the boom shall be lowered below the point where less than two full wraps of rope remain on their respective drums.

Holding the Load

The operator shall not be allowed to leave his/her position at the controls while the load is suspended. If the load must remain suspended for any considerable length of time, a dog, or pawl and ratchet, or other equivalent means, rather than the brake alone, shall be used to hold the load.

Securing Boom

Dogs, pawls, or other positive holding mechanism on the hoist must be engaged. When not in use, the derrick boom shall be:

- Laid down.
- Secured to a stationary member, as nearly under the head as possible, by attachment of a sling to the load block.
- Hoisted to a vertical position and secured to the mast.

OTHER REQUIREMENTS

Guards

Exposed moving parts, such as gears, ropes, setscrews, chains, chain sprockets, and reciprocating components, which constitute a hazard under normal operating conditions, shall be guarded.

Operating Near Electrical Power Lines

Except where electrical distribution and transmission lines have been deenergized and are visibly grounded at the point of work, or when insulation barriers, not part of the derrick, have been erected to prevent physical contact with the lines, the minimum clearance between the lines and any part of the derrick or load shall be:

- For lines rated 50 kV or below, ten feet.
- For lines rated over 50 kV, ten feet plus 0.4 inch for every kV over 50 kV, or twice the length of the line insulator but never less than ten feet.

Notification

Before commencement of operations near electrical lines, owners of the lines, or their authorized representatives, shall be notified and provided with all pertinent information.

Overhead Wire

Any overhead wire shall be considered to be an energized line unless and until the owner of the line or the electrical utility authorities indicate that it is not an energized line.

CONSTRUCTION

The applicable OSHA construction regulation which applies to derricks is found in 29 CFR 1926.550 which pertains to cranes and derricks.

Derricks

All derricks in use meet the applicable requirements for design, construction, installation, inspection, testing, maintenance, and operation as prescribed in American National Standards Institute B30.6-1969, Safety Code for Derricks. For floating cranes and derricks, when a mobile crane is mounted on a barge, the rated load of the crane may not exceed the original capacity specified by the manufacturer.

A load rating chart, with clearly legible letters and figures, is provided with each crane by the manufacturer, and are securely fixed at a location easily visible to the operator. When load ratings are reduced, for example to stay within the limits listed for a barge with a crane mounted on it, a new load rating chart is to be provided.

Floating Cranes and Derrick

Mobile cranes on barges must be positively secured. For permanently mounted floating cranes and derricks, when cranes and derricks are permanently installed on a barge, the capacity and limitations of use are based on competent design criteria. A load rating chart, with clearly legible letters and figures, is to be provided and securely fixed at a location easily visible to the operator. Floating cranes and floating derricks in use must meet the applicable requirements for design, construction, installation, testing, maintenance, and operation as prescribed by the manufacturer. The employer shall comply with the applicable requirements for protection of employees working onboard marine vessels.

REFERENCES

Reese, C. D. and J.V. Eidson. *Handbook of OSHA Construction Safety & Health.* Boca Raton, FL: CRC/Lewis Publishers, 1999.

United States Department of Labor. Occupational Safety and Health Administration. OSHA 10 and 30 Hour *Construction Safety and Health Outreach Training Manual.* Washington, DC, 1991.

United States Department of Labor. Occupational Safety and Health Administration. General Industry. Code of Federal Regulations. Title 29, Part 1910. Washington, GPO, 1998.

United States Department of Labor. Occupational Safety and Health Administration. Construction. *Code of Federal Regulations.* Title 29, Part 1926. Washington, GPO, 1998.

United States Department of Labor. Occupational Safety and Health Administration. *OSHA Technical Manual.* Chapter 13—Cranes and Derricks. Washington, DC, 1990.

United States Department of Labor. Occupational Safety and Health Administration. Office of Training and Education. OSHA *Voluntary Compliance Outreach Program: Instructors Reference Manual.* Des Plaines, IL, 1993.

CHAPTER 7

HELICOPTERS

A K-MAX operated by ROTEX AG of Liechtenstein positioning and affixing cable guides for a ski lift in the German Alps. Permission by the Kaman Corporation

Regulations for the use of helicopters are found in 29 CFR 1910.183 and 29 CFR 1926.551. These OSHA safety and health regulations are the same for both the general industry and construction industry. The intent of these regulations is to insure that ground crews and flight crews are aware of safety hazards and have made every effort to protect the safety of all concerned. Safety measures must be understood by all individuals involved with the use of helicopters for lifting on worksites. The following paragraphs summarize the content of these regulations.

The use of helicopters for lifting requires special procedures, an adequately trained ground crew, and an awareness of the unique hazards involved with their use. Helicopter cranes are expected to comply with all applicable regulations of the Federal Aviation Administration. Prior to each day's operation, a briefing is to be conducted. This briefing must set forth the plan of operation for the pilot and ground personnel.

Every load must be properly slung. Tag lines should be of a length that will not be drawn up into rotors. Pressed sleeves, swedged eyes, or equivalent means are to be used for all freely suspended loads in order to prevent hand splices from spinning open, or cable clamps from loosening.

All electrically operated cargo hooks must have electronically activated devices designed and installed to prevent inadvertent operation. Cargo hooks must be equipped with an emergency mechanical control for releasing the load. Each hook is to be tested prior to each day's operation to determine that the release functions properly, both electrically and mechanically.

Workers receiving the load must have personal protective equipment, including complete eye protection and hard hats secured by chinstraps. Loose-fitting clothing, which is likely to flap in the downwash and be snagged on a hoist line, should not be worn. Every practical precaution shall be taken to protect the employees from flying objects in the rotor downwash. Within 100 feet of the place where the load will be lifted or deposited, and areas susceptible to rotor downwash, all loose gear must be secured or removed. Good housekeeping is to be maintained in helicopter loading and unloading areas. The helicopter operator or pilot shall be responsible for the size, weight, and manner in which loads are connected to the helicopter. Open fires are not permitted in areas where fires can be spread by rotor downwash. If, for any reason, the helicopter operator believes the lift cannot be made safely, the lift should not be made. The weight of an external load must not exceed the manufacturer's rating.

When employees are required to perform work under hovering craft, a safe means of access must be provided for employees to reach the hoist line hook, and engage or disengage cargo slings. Employees shall not perform work under hovering craft except when necessary to hook or unhook loads. (See Figure 7-1.)

Static electrical charge on the suspended load is to be dissipated with a grounding device before ground personnel touch the suspended load. Also, electrical protective rubber gloves must be worn by all ground personnel touching the suspended load.

Ground lines, hoist wires, or other gear (except for pulling lines and conductors) which is allowed to "pay out" from a container or roll off a reel is not to be attached to any fixed ground structure or allowed to foul on any fixed structure.

When visibility is reduced by dust or other conditions, ground personnel must exercise special caution to keep clear of main and stabilizing rotors. Precautions must also be taken by the employer to eliminate, within what is practical, the causes of reduced visibility. Signal systems between aircrew and ground personnel must be understood and checked in advance of hoisting the load. This applies to either radio or hand signal systems. Hand signals are as shown in Figure 7-2.

These regulations (29 CFR 1926.551 and 29 CFR 1910.183) include the following:

- Helicopter regulations —Helicopter cranes shall comply with any applicable regulations of the Federal Aviation Administration. (See Figure 7-3.)

- Briefing—Prior to each day's operation, a briefing must be conducted that sets forth the plan of operation for the pilot and ground personnel.

- Slings and tag lines—Loads shall be properly slung. Tag lines shall be of a length that will not permit their being drawn up into the rotors.

- Cargo hooks —All electrically operated cargo hooks must have an electronically activated device so designed and installed as to prevent inadvertent operation. In

Figure 7-1. A ROTEX K-MAX helicopter hauls structural components of a hammerhead tower crane into place. Permission by the Kaman Corporation

addition, these cargo hooks shall be equipped with an emergency mechanical control for releasing the load. The cargo hooks must be tested prior to each day's operation by a competent person to determine that the release functions properly, both electrically and mechanically.

- Personal protective equipment — Personal protective equipment (complete eye protection and hard hats secured by chinstraps) shall be provided by the employer

Figure 7-2. Hand Signals Used by Helicopter Ground Personnel. Courtesy of OSHA from Figure N-1 of 29 CFR 1926.551 and 29 CFR 1910.183

and used by employees receiving the load. Loose-fitting clothing likely to flap in rotor downwash, and thus be snagged on the hoist lines, may not be worn.

- Loose gear and housekeeping—All loose gear within 100 feet of the place of lifting the load or depositing the load, or within all other areas susceptible to rotor downwash, must be secured or removed. Good housekeeping shall be maintained in all helicopter loading and unloading areas.

- Hooking and unhooking loads—Employees are not permitted to perform work under hovering craft except when necessary to hook or unhook loads.

- Static charge—Static charge on the suspended load shall be dissipated with a grounding device before ground personnel touch the suspended load, unless protective rubber gloves are being worn by all ground personnel who may be required to touch the suspended load.

- Signal systems—The employer shall instruct the air crew and ground personnel on the signal systems to be used and shall review the systems with the employees before hoisting the load. This applies to both radio and hand signal systems. Hand signals, where used, are to be in conformance with Figure 7-2.

- Approach distance—No employee shall be permitted to approach within 50 feet of the helicopter when the rotor blades are turning, unless their work duties require their presence in that area.

- Communications—There shall be constant reliable communication between the pilot and a designated employee of the ground crew who acts as a signalman during the period of loading and unloading. The signalman must be clearly distinguishable from other ground personnel.

Figure 7-3. K-MAX performing a Vertrep mission (vertical replenishment) of ships a sea for the U.S. Navy. Permission by the Kaman Corporation

REFERENCES

Reese, C. D. and J.V. Eidson. *Handbook of OSHA Construction Safety & Health.* Boca Raton, FL: CRC/Lewis Publishers, 1999.

United States Department of Defense. *Multiservice Helicopter Sling Load: Basic Operations and Equipment.* Washington, DC, 1997.

United States Department of Labor. Occupational Safety and Health Administration. *OSHA 10 and 30 Hour Construction Safety and Health Outreach Training Manual.* Washington, DC, 1991.

United States Department of Labor. Occupational Safety and Health Administration. General Industry. *Code of Federal Regulations.* Title 29, Part 1910. Washington, GPO, 1998.

United States Department of Labor. Occupational Safety and Health Administration. Construction. *Code of Federal Regulations.* Title 29, Part 1926. Washington, GPO, 1998.

United States Department of Labor. Occupational Safety and Health Administration. *OSHA Technical Manual.* Chapter 13—Cranes and Derricks. Washington, DC, 1990.

United States Department of Labor. Occupational Safety and Health Administration. Office of Training and Education. *OSHA Voluntary Compliance Outreach Program: Instructors Reference Manual.* Des Plaines, IL, 1993.

CHAPTER 8

HOISTS

Typical hoist used for lifting

Hoists, in some form or fashion, have been utilized for hundreds of years. Hoists are used to lift and lower heavy loads ranging from thousands of pounds to thousands of tons. Suffice it to say that hoists are an integral part of many material handling activities. The OSHA General Industry standards relevant to hoists and hoisting equipment are found in 29 CFR 1910.179 (h).

GENERAL GUIDELINES

The following are established safety guidelines for the operation, inspection, testing, and design of electric-, air-, or hand-powered hoists, which are not permanently mounted on overhead cranes. ANSI/ASME B30.16 prescribes good work practices. A written copy should be affixed to the hoist control or nearby to reinforce proven practices for safe hoist operations. An example of safe operating procedures can be found in Figure 8-1.

HOIST—SAFE OPERATING PROCEDURES

SAFETY PRECAUTIONS

AT THE BEGINNING OF EACH SHIFT OF USE

- Operate each control to determine if it functions properly under no load. Evaluate lubrication and other service needs by observing and listening to the equipment.
- Visually inspect chains, rope, slings, and hooks before lifting a load. Have worn, cracked, frayed or otherwise damaged components replaced before use.
- Test the braking system by lifting the load a few inches from the working surface and suspending it by the brake.

DURING USE

- Loads should be well secured before lifting and slings should be adequately arranged for the load.
- All persons should be clear before the lift of the load begins.
- Avoid sudden stops or starts when moving loads laterally.

PARKING THE HOIST

- The hoist block should be positioned above head level when the hoist is not in use.

IF LYING SYSTEM APPEARS UN-SAFE INSTALL "OUT-OF-ORDER" TAG AND REPORT TO SUPERVISION IMMEDIATELY

(FRONT)

WARNING!!!

- Do not operate this hoist/crane unless you are authorized by area supervision.
- Do not lift a load which is greater than the weight capacity of the hoist or other lifting equipment.
- Do not operate hoist with twisted, kinked, or damaged chain or wire rope.
- Do not operate a hoist with a wire rope that is not properly seated in its groove.
- Do not operate a damaged or malfunctioning hoist.
- Do not lift people or loads over people.
- The hoist must be centered over the load before lifting.
- Do not leave the controls while the load is suspended.
- Do not use power to operate a manual hoist.
- Do not remove or obscure this tag.

QUESTIONS REGARDING THE SAFE OPERATION OF THIS HOIST/ CRANE SHOULD BE ADDRESSED IMMEDIATELY WITH SUPERVISION.

(BACK)

Figure 8-1. Safe operating procedures for a hoist

Operators of hoists must be trained and qualified in accordance with regulatory requirements. The special-rated load capacity shall be permanently marked on the hoist or load block and must be clearly legible from the operating position. The hoist shall be permanently marked with the manufacturer's name, address, and unit identification.

Load chains are required to be proof tested by the chain or hoist manufacturer with a load at least one-and-one-half times the rated load, divided by the number of chains supporting the load. Wrought iron chain is not permitted for load line. Roller chain is also not permitted for load lines.

The hoist must be designed so that, when the actuating force is removed, it will automatically stop and hold the load up to 125 percent of the special rated load capacity. The special rated load capacity may never be exceeded except for properly authorized tests. The hoist shall be installed only in locations that will permit the operator to stand free of the load at all times. Support structures, including trolleys, monorails, and cranes, if any, shall have a special-rated capacity at least equal to that of the hoist.

Both safety features and operation must conform, at least to the minimum, to the provisions of ANSI B30.16, and electrically powered chain hoists shall comply with ANSI/ASME HST-1M.

INSPECTIONS

All new hoists shall be inspected by the hoist manufacturer. Prior to initial use, all altered or repaired hoists shall be inspected by, or under the direction of, a certified inspector whose qualifications to perform specific activities has been determined, verified, and attested to in writing to ensure compliance with these provisions. Figure 8-2 is a safety and health checklist for a hoist.

Inspection procedures for hoists in regular use are divided into two general classifications, based upon the intervals at which the inspection should be performed by or under the direction of a certified inspector whose qualifications have been determined, verified and attested to in writing. The intervals, in turn, are dependent upon the nature of the critical components of the hoist, and the degree of their exposure to wear, deterioration, or malfunction. General classifications are designated as "frequent" and "periodic," as defined below:

1. Frequent inspection: daily to monthly intervals

2. Periodic inspection: 1- to 12-month intervals, or as specified by the manufacturer. Hoists exposed to adverse environments should receive periodic inspections more frequently.

Frequent Inspections

The "frequent" classification requires that hoists must be inspected for damage at defined intervals or intervals as otherwise established. Inspection must include observations during operation to detect any damage, which might appear between regular inspections. Deficiencies shall be carefully examined, and determinations made as to whether they constitute safety hazards. Inspections should include:

1. All controls and operating mechanisms for improper operation: daily or before use.

2. All safety devices for malfunction: daily or before use.

3. Deterioration or leakage in air systems: daily or before use.

Safety and Health Checklist for Hoists

_____ Are all hoists inspected by a designated person to insure that the special rated load capacity will not be exceeded, and to be sure that the hoist are still within the test-certification period?

_____ Are all lifting devices inspected by a designated person to make certain that the special rated load capacity of each item device is not exceeded, and that they are still within the test certification period, if appropriate?

_____ Has the operator, or other designated person, visually examine the hoist (records are not required except as noted) in accordance with the requirements for a frequent inspection?

_____ Have any deficiencies been carefully examined by a qualified person, to determine whether they constitute a hazard? Have these deficiencies been corrected prior to operating the hoist?

_____ Have load lines been checked after strain is put on them, and before the load is lifted clear of the ground? If not plumb, have the slings or equipment been repositioned so that the lines are plumb before continuing?

_____ Is the safe load capacity posted on the body of the hoist?

_____ Are there legible safe operating procedures attached to the hook, block, hoist, or controls?

_____ Is the hoist securely attached to its supports?

_____ If a hoist is secured by hooks to its support, does hook have latches?

_____ Can hoist supports bear the maximum intended load?

_____ Are the positive stops or limiting devices for the hoist on the rails, tracks, or trolleys?

_____ Is the maximum load capacity posted on the overhead rails?

_____ Are flanges and spiral grooves on hoist drums free of projections?

_____ Is the hoist positioned directly over the load by the operator?

_____ Have all personnel been forbidden to be under a raised load?

_____ Are operators forbidden to lift or transport personnel?

_____ Has the hoist been inspected prior to use for wear, damage, or malfunction of any of its components?

_____ Are controls designed and functioning properly so that when released by the operator, they will stop and not let the load fall?

Figure 8- 2. Safety and health checklist for a hoist

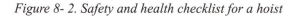

4. Hooks damaged from chemicals, deformations, cracks, or having more than 15 percent in excess of normal throat opening, or more than a 10-degree twist from the plane of the unbent hook. Note that any hook that is twisted or has a throat opening in excess of normal indicates abuse or overloading of the unit. Other load bearing components of the hoist should be inspected for damage: daily or before use.

5. Load-carrying ropes or chains: visual inspection shall be made daily for wear, twist, distortion, or improper dead-ending to the hoisting drum and other at-

tachments. Chains must also be inspected for deposits of foreign material, which may be carried into the hoist mechanism.

Periodic Inspections

The "periodic" classification calls for a complete inspection of the hoist at intervals as previously described, depending upon the unit's activity, severity of service, and environment, or as specifically indicated below. These inspections shall include the requirements of "frequent" inspections above plus items such as those listed below. Any deficiencies found during inspection must be carefully examined, and determinations shall be made as to whether they constitute a safety hazard. Inspection records must kept on file and readily available. Items for inspection include:

1. Loose bolts or rivets.

2. Cracked or warn drums or sheaves.

3. Worn, corroded, cracked, or distorted parts such as pins, bearings, shafts, gears, rollers locking, and clamping devices.

4. Excessive wear on motors or load brakes.

5. Excessive wear of chains, ropes, load sprockets, drums, sheaves, and chain stretch.

Semiannual inspections are to be completed with written, dated, and signed reports, and must be conducted on critical items, i.e., brakes, hooks, chains, and load lines. Inspection records shall be kept on file and readily available. Item on which to focus during a semiannual inspection are:

1. Hooks: inspection may involve dye penetrants, magnetic particles, or other suitable crack-detecting procedures. Inspections should be performed at least once per year.

2. Electrical apparatus, for signs of pitting or any deterioration, of controller contractor limit switches, and push button switches.

3. Hook retaining nuts or collars and pins, welds, or riveting used to secure the retaining members.

4. Supporting structures and trolleys, if used, shall be inspected for continued ability to support the imposed loads.

5. Warning labels for absence or illegibility.

Rope Inspections

All running ropes shall be visually inspected by the operator or an appointed person once each shift or prior to use if the hoist is not in regular service. A thorough inspection of all ropes shall be made by a qualified inspector at least once every six months, or prior to use if the hoist is not in regular service. Written, dated, and signed inspection reports shall be kept on file and readily available to appointed personnel. Any deterioration resulting in appreciable loss of original strength such as described below shall be carefully noted, and determinations must be made as to whether further use of the rope constitutes an acceptable risk.

1. Reduction of rope diameter below nominal, whether due to loss of core support, internal or external corrosion, or wear of outside wires.

2. A number of broken outside wires and the degree, distribution, or concentration of such broken wires.

3.   Worn outside wires.

4.   Sections of rope which are normally hidden during inspection or maintenance procedures, such as parts passing over sheaves, should be given close inspection, as these are points most subject to deterioration.

5.   Corroded or broken wires at end connections.

6.   Corroded, cracked, bent, worn or improperly applied end connections.

7.   Kinking, crushing, cutting, or unstranding.

All rope which has been idle for a month or more due to shutdown or storage of the hoist on which it is installed must be given a thorough inspection before it is returned to service. The inspection shall be for all types of deterioration, and shall be performed by an appointed or designated person whose approval is required before further use of the rope. A written and dated report of the rope's condition must be filed.

No precise rules can be given for determining the exact interval for replacement of ropes, since many variables are involved. After allowance for deterioration disclosed by inspection, safety in this respect depends largely upon the use of good judgment by an appointed or designated person who evaluates the remaining strength in a used rope. In order to establish data as a basis for judging the proper time for replacement, an inspection record shall be maintained. Conditions such as the following shall be sufficient reason for questioning rope safety and for considering its replacement:

1.   In hoist ropes, 12 randomly distributed broken wires in one rope lay, or three broken wires in one strand in one rope lay.

2.   Wear of one-third of the original diameter of outside individual wires.

3.   Kinking, crushing, birdcaging, or any other damage resulting in distortion of the rope structure.

4.   Evidence of any heat damage from any cause.

5.   Reductions from nominal diameter of more than 0.4 millimeter (mm) (1/64 inch) for diameters up to and including 7.9 mm (5/16 inch), 0.8 mm (1/32 inch) for diameters 9.5 mm (3/8 inch) up to and including 12.7 mm (1/2 inch), 1.2 mm (3/64 inch) for diameters 14.3 mm (9/16 inch) up to and including 19.1 mm (3/4 inch), 1.6 mm (1/16 inch) for diameters 22.2 mm (7/8 inch) up to and including 28.6 mm (11/8 inch), and 2.4 mm (3/32 inch) for diameters 31.8 mm (11/4 inch) up to and including 38.1 mm (11/2 inch).

Special attention must be given to end fasteners. Ropes should be examined frequently at socketed fittings. Upon the development of two broken wires adjacent to this point, the rope should be socketed or replaced. Resocketing shall not be attempted if the resulting rope length will be insufficient for proper operation. Portions of rope subject to reverse bends and rope that operates over small-diameter drums and sheaves should be given close attention.

Any replacement rope shall be the same size, grade, and construction as the original rope furnished by the hoist manufacturer, unless otherwise recommended by a rope manufacturer to safely handle specific working conditions.

Chain Inspections

Operate the hoist under load in hoisting and lowering directions and observe the operation of the chain and sprockets. The chain should feed smoothly into and away from the sprockets. If the chain binds, jumps, or is noisy, first see that it is clean and properly lubricated. If the trouble persists, inspect the chain and mating parts for wear, distortion, or other damage.

The chain should be cleaned prior to inspection. Examine it visually for gouges, nicks, weld splatter, corrosion, and distorted links. Slacken the chain and move adjacent links to one side to inspect for wear at the contact points. If wear is observed or stretching is suspected, the chain should be measured according to the hoist manufacturer's instructions. If instructions are not available, proceed as follows. Select an unworn, unstretched length of the chain (e.g., at the slack end) and suspend the chain vertically under tension. Using a caliper-type gauge, measure the outside length of any convenient number of links in an approximately 12 to 14 inch section. Measure the same number of links in the used sections and calculate the percentage increase in length. If the used chain exceeds a hoist manufacturer's recommended length, or in the absence of such a recommendation, if the used chain is one-and-a-half percent longer than unused chain, replace the chain. Repairing the load chain, by welding or any other means, shall not be attempted by anyone other than the chain manufacturer.

The existence of gouges, nicks, corrosion, weld splatter, or distorted links is sufficient reason for questioning chain safety and considering chain replacement. Safety in this respect depends largely upon the use of good judgment by an appointed or designated person in evaluating the degree of damage. Replacement chain shall be the same size, grade, and construction as the original chain furnished by the hoist manufacturer, unless otherwise recommended by the hoist manufacturer based upon the actual working condition.

Load chain links which pass over the hoist-load sprocket on edge (rather than those that lie flat in the pockets) should be installed with the welds away from the center of the sprocket. (This precaution is not required on idler sprockets, which change the direction but not the tension in the chain.) The chain shall be installed, without any twist, between the hoist and an anchored end on either the loaded side or the slack side. When a chain is replaced, disassemble and inspect the mating parts (e.g., sprockets, guides, and stripper) for wear. Replace the chain if necessary.

LOAD TESTING

All hoists in which load-sustaining parts which have been altered, replaced, or repaired shall be tested and inspected by a qualified inspector and a written report confirming the load rating shall be filed and readily available. Test loads shall be at 125 percent of the rated capacity. (See Figure 8-3 for inspection and load test form.) On hoists incorporating overload devices that prevent the lifting of 125 percent of the rated load, a load test shall be accomplished with at least 100 percent of this rated load, followed by a test of the function of the overload device.

The trip setting of limit switches and limiting devices shall be inspected first by hand, if practical, then under the slowest speed obtainable. Continue the test with increasing speeds up to the maximum. Actuating mechanisms shall be located so that they will trip the switches or limiting devices in sufficient time to stop motion without damaging any part of the hoisting arrangement. All anchorages and/or suspensions must be approved by a designated inspector to ensure compliance with the general industry and construction standards, and tests must include the following functions: hoisting and lowering, operation of brakes, limit devices, locking, and safety devices.

Permanently installed hoists shall be load tested when first assigned to service, thereafter at three year intervals, and when specified in the employer's routine and normal inspection schedule. Except when temporarily installed, portable units (i.e., chain falls) shall be tested

LOAD TEST AND INSPECTION FORM

INSPECTED BY:_____ DATE:_____

HOIST ID#_____ LOCATION:_____

NOTES: 1. Load test at 100 percent of manufacturer's rated capacity/load test shall not exceed 100 percent of manufacturer's rated capacity. Test weights shall be accurate to within five percent plus zero percent of stipulated values.

2. Inspector shall verify all steps as listed below.

3. Craftsmen will initial all tests, work, and inspections completed below.

4. All inspections shall be in accordance with OSHA regulations.

---

1. Perform the annual preventive maintenance inspection: Check unit for proper operation.

2. Manual hoists only: Check brake mechanism for worn, glazed, or contaminated disks, worn pawls, cams, or ratchets. Check for broken, corroded, or stretched pawl springs. Repair as needed.

3. Electrical powered hoist: Check for:

   a. All functional operating mechanisms for misadjustment interfering with proper operation.

   b. Limit switches or devices for proper operation.

   c. External evidence of damage or excessive wear of load sprockets, idler sprockets, and drums or sheaves.

   d. External evidence of wear on motor or load brake.

   e. Electrical apparatus for signs of pitting or any deterioration of visible controller contacts.

   f. All anchorage or hoist suspensions.

4. Set hoist up for load test and inspection. Where applicable insure that the load chart is legible.

5. Perform load test using the required test weights and appropriate slings. Measure a length of the load chain under tension; measure a length of 12 to 14 links. If wire rope is used measure the diameter. Inspect chain and/or wire rope in accordance to OSHA regulations.

If Hoist Is Equipped With a Trolley:

1. Mount hoist on a monorail.

2. Rig test weight to load hook.

Figure 8-3. Inspection and load test form. Courtesy of the U.S. Department of Energy

3.   Perform load test moving weight along monorail. Observe hoist and trolley. Observe performance of load bearing components.

4.   Lower test weight to floor. Note performance of hoist during lowering operation; remove rigging.

At the completion of the load test, inspect the following items.

1.   Visually inspect and remeasure the load chain and/or hoist rope after the load test. Check fordeformed or broken links, stretch, and etc.

2.   Inspect load hook and suspension hook for bend or twist.

<u>LOAD HOOK</u>:   <u>PREVIOUS</u>   <u>PRESENT</u>

Inspector Verify_____ Throat Opening: _____ _____

Inspector Verify_____ Hook Twist: _____ _____

<u>SUSPENSION HOOK</u>:   <u>PREVIOUS</u>   <u>PRESENT</u>

Inspector Verify_____ Throat Opening: _____ _____

Inspector Verify_____ Hook Twist: _____ _____

Inspector perform NDT test on hook by visual examination, liquid penetrant examination, or magnetic particles examination. Acceptance: No cracks, linear indications, laps, or seams.

Hooks with more than 15 percent normal (new hook) throat opening shall be replaced. Hooks with more than ten degrees twist from the normal (new hook) plane of the hook, shall be replaced. Hooks having more than ten percent wear in the throat section or five percent elongation of the shank shall be replaced. Lubricate hooks at each inspection.

Establish three marks, A, B, and C with a center punch. For ease in measuring, set distances on an even number of inches.

<u>BEFORE LOAD TEST</u> *   <u>AFTER LOAD TEST</u>*

Length AB_____ in.   Length AB _____ in.

Length BC_____ in.   Length BC _____ in.

See diagram below:

Check for:

•   Wear and deformation

•   Cracks and twisting

•   Signs of opening between Point A and Point B

Equipment Operator:_____

Actual Load Test _____lbs. Quality Verify Load Test_____ Date _____

Figure 8-3. (Continued) Inspection and load test form. Courtesy of the U.S. Department of Energy

annually. In no case shall the load test exceed the rated capacity of the equipment. Test weights shall be accurate to within five percent plus zero percent of stipulated values. Load tests shall be performed by a qualified inspector and load test records shall be kept on file and readily available.

If a test has not been completed by the end of the required period, the equipment must be downrated as follows:

1.　Thirty calendar days after the end of the period, the equipment shall be downrated to 75 percent of the rated capacity.

2.　Sixty calendar days after the end of the period, the equipment shall be downrated to 50 percent of the rated capacity.

3.　Ninety calendar days after the end of the period, the equipment shall be taken out of service until the required inspection has been completed.

High-Consequence Lifts

All previous requirements must be met and also include the following:

1.　Static Test: Equipment shall hold the test load for ten minutes, or the time required to check all primary load bearing parts while under strain without slippage, damage, or permanent deformation of any part of the equipment. Hoisting equipment and winches shall be tested in maximum run-out of the hoisting rope or chain, when practical.

2.　Dynamic Test: Hoisting equipment shall be operated through at least two complete cycles of all movements that the equipment will encounter in service, while supporting the test load. At a minimum, the test load shall be raised far enough for all drums, sheaves, gears, and other rotating parts of the hoisting mechanism to complete at least one or, if possible in the clearance available, two complete revolutions; then lowered until the load comes within two or three inches of the ground; and held at this level for one minute. This hoisting cycle must be repeated at least one more time. Tests shall be made by the operator who normally operate with this equipment. He or she should demonstrate the ability to positively control the load during all lateral, rotational, and vertical motions that will be encountered in service. At least once during the lifting portion of the hoisting cycle, and once during the lowering cycle, power to the hoisting equipment shall be completely shut off. There shall be no slippage of the load or overheating of the brakes.

OPERATING PRACTICES

The operator shall not engage in any practice which will divert his or her attention while engaged in operating the hoist. Before starting the hoist, the operator must be certain that all personnel are clear of the equipment. The operator shall familiarize himself or herself with the equipment and its proper care. If adjustments or repairs are necessary or any damage is known or suspected, the operator is obligated to report it promptly to the appointed person. The operator shall also notify the next operator of the problem upon changing shifts. All controls must be tested by the operator before beginning a shift. If any controls do not operate properly, they shall be adjusted or repaired prior to starting operations First and foremost, make sure hands are clear from all moving parts. Take signals from only one person using the standard signs. (See Figure 8-4.) A stop signal shall be obeyed regardless of who gives it.

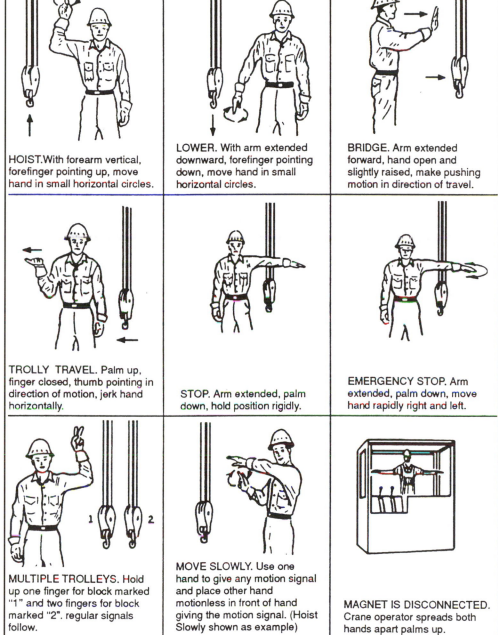

Figure 8-4. Standard hand signals for a hoist and overhead/gantry cranes. Permission by the American Society of Mechanical Engineers

CONTROLS FOR THE HOIST

Controls should be designed so that when they are released by the operator, the hoist will come to a stop and not release the load. Control levers which move up should raise the load, but when released, should be spring-loaded so that they return to a neutral position stopping or holding the load. Controls should perform the same for down movement. When actuator buttons are used, they should be spring-loaded. The top button should be for raising a load, and the bottom button for lowering a load. It would be best if the buttons and levers were marked with arrows indicating the direction of movement coinciding with the movement of the load lifted by the hoist.

MOVING A LOAD

The hoist shall not be loaded beyond the special rated load capacity, except for authorized testing. The hoisting rope or chain must not be wrapped around the load. The load should be attached to the hook by means of slings or other approved devices. The slings or other approved devices shall be seated properly in the saddle of the hook before hoisting operations are carried out.

The load must not be moved or lifted more than a few inches until it is well balanced in a sling or lifting device. Care should be taken in hoisting to be certain that:

1.   The hoist rope or chain is not kinked or twisted.

2.   The load does not contact any obstructions.

3.   Multiple-part ropes or chains are not twisted around each other.

Before starting to hoist, the rope or chain shall be properly seated on the drum, sheaves, or sprockets. Hoists shall not be operated until the hoist unit is centered over the load, except when specifically authorized by an appointed person who has determined that the components of the hoist and its mounting will not be overstressed. A hoist should never be used for handling personnel unless specifically recommended for such use by the manufacturer, and so indicated on a permanent name plate attached to the hoist. Equally important, the operator must avoid carrying loads above people.

The operator shall test the brakes each time a load approaching the rated load capacity is handled. Raise the load just enough to clear the floor or supports, and check for brake action. Continued the lift only after the operator is certain that the braking system is operating properly.

No loaded hoist drum shall be rotated in the lowering direction beyond the point where less than two wraps of rope remain in the drum. Also, the operator shall inch the hoist into engagement with a load, and avoid unnecessary stops and starts. A tag line shall be used. If it becomes necessary, a tagline shall be used to guide, snub, or otherwise control the load.

HOIST-LIMIT SWITCH

At the beginning of a shift, the operator shall try out the upper-limit switch of each hoist under no load. Extreme care should be exercised as the block is "inched" into the limit, or run at slow speed. If the switch does not operate properly, the designated person shall be notified immediately. If a lift is in progress during a shift change, this testing requirement is considered to have been satisfied for the completion of that lift. However, the limit switch must be tested again before the next lift. A hoist limit switch that controls the upper limit travel of the load block shall never be used as an operating control.

SUGGESTED LIFTING PROCEDURES

The person in charge (PIC) make sure that a work plan is prepared covering the entire lifting operation. Consideration of the lift history of the hoist shall affect the plan, which shall also include, but is not limited to, sling angles and sizes; inspection and test-certification periods; load configuration; the presence of hazardous materials; and the requirement for a load indicating device. The plan must be reviewed and approved by the cognizant safety organization.

When the weight of the lift is within ten percent of the special rated load capacity, and equipment of greater capacity is not available, the PIC shall review, in detail, the positioning and rigging of the load with the person who will carry out the lift. The effects of ground conditions, wind and weather on the stability of the equipment, and the effect of rotational and translational speeds shall be considered in giving instructions to equipment operators.

Hoisting shall be stopped when the load is located approximately two inches from the supports. A check must be made to determine any potential for the load to swing or sway, and any possibility that slings may slip or change position. Sling positions shall be adjusted, and additional supports or restraints shall be added as necessary before continuing the lift. Approval to change position of the sling position, supports, and restraints shall be obtained from the PIC before procedures may be modified.

The PIC shall be a qualified person, or shall be assisted by a qualified person, experienced in using hoisting and rigging equipment of the type in use. He or she must prepare and review drawings, procedures, and equipment assignments, and supervise the job.

General procedures may be submitted for handling loads, except when detailed procedures are specified. Individual procedures for several similar processes are not required unless specified. Procedures must include identification of the equipment or class of equipment to be used, their special rated load capacities, any special instructions to operators, and provisions for verification by the authorized person that the lift or move has been satisfactorily completed. The PIC shall certify that good rigging precautions, practices, and safety measures will be employed; that equipment operators are qualified and have been properly instructed; and that equipment is adequate for the loads involved and in good operating condition.

In addition, the work plan should include the following:

1. Identification of each piece of operating equipment to be used in the move by type, rated and special rated load capacity and, for other than permanently installed equipment, the equipment serial number or other identifying number. (If the specific piece of nonpermanent equipment has not been identified at the time of procedure specification, the number shall be included in the job instructions prepared by the appointed person.)

2. Identification of slings, lifting bars, and other major rigging accessories or assemblies by serial number. (If the specific equipment has not been identified at the time of procedure preparation, it shall be identified in the job instructions prepared by the appointed person.)

3. A list of all nonserialized rigging accessories and materials required in the move or lift, identified by type and capacity, should be developed.

4. Identification of the item to be moved, its weight, dimensions, and center of gravity (as determined by the method of SAE J874, or estimated from drawings or engineering analysis), and the total hook load.

5. Rigging sketches showing all lifting points, load vectors, sling angles, accessories, methods of attachment, and other factors affecting the capacity of equipment and limitations to be applied.

6. Approximate and maximum hoist and winching speeds.

7. Instructions to be given to operators, including the sequence of equipment moves and coordination with moves of other equipment involved; translational speeds, directions, and distances; load weight and center of gravity; and other pertinent data.

8   Identification of persons who will have field responsibility for the move or lift and for monitoring it.

9, Requirements for specific tests to be made before, during, and after the move or lift, including load tests for high consequence loads or practice lifts.

10. Surveillance procedures, including check points, instruments, and indicators that will be used to ensure that the move proceeds according to plan, and the special-rated capacity of the equipment is not exceeded.

11. Provisions for verification by the PIC, or his/her designee, of satisfactory completion of each step of the procedure as it occurs.

BASE-MOUNTED DRUM HOIST (CFR 1926.553)

Base-mounted drum hoists with exposed moving parts, such as gears, projecting screws, setscrews, chains, cables, chain sprockets, and reciprocating or rotating parts, which constitute a hazard, must be guarded. All controls used during the normal operation cycle are to be located within easy reach of the operator's station. Electric motor-operated hoists are to be provided with a device to disconnect all motors from the line upon power failure, and no motor is permitted to be restarted until the controller handle is brought to the "off" position. And, where applicable, an overspeed preventive device is to present. Also, a means must be present whereby a remotely operated hoist-stop can be activated when any control is ineffective. All base-mounted drum hoists in use must meet the applicable requirements for design, construction, installation, testing, inspection, maintenance, and operations, as prescribed by the manufacturer.

MATERIALS HOIST (1926.552)

All material hoists must comply with the manufacturers' specifications and limitations, which are applicable to the operation of all hoists and elevators. Where manufacturers' specifications are not available, limitations assigned to the equipment are to be based upon the determinations of a professional engineer competent in the field. Rated load capacities, recommended operating speeds, and special hazard warnings or instructions are to be posted on cars and platforms.

Hoisting ropes are to be installed in accordance with the wire rope manufacturers' recommendations. Wire rope is to be removed from service when any of the following conditions exists.

1. In hoisting ropes, six randomly distributed broken wires in one rope lay, or three broken wires in one strand in one rope lay.

2. Abrasion, scrubbing, flattening, or peening, causing loss of more than one-third of the original diameter of the outside wires.

3. Evidence of any heat damage resulting from a torch, or any damage caused by contact with electrical wires.

4. Reduction from nominal diameter of more than 3/64 inch for diameters up to and including 3/4 inch; 1/16 for diameters 7/8 to 1 1/8 inches; and three thirty-seconds inch for diameters 1 1/4 to 1 1/2 inches.

The installation of live booms on hoists is prohibited. The use of endless belt-type manlifts on construction is prohibited. Operating rules for material hoists are to be established and posted at the operator's station of the hoist. Such rules include signal systems and the allowable line speed for various loads. Rules and notices are to be posted on the car frame or crosshead in a conspicuous location. These rules must include the statement, "No Riders Allowed," since no person is allowed to ride on material hoists except for the purpose of inspection and maintenance.

All entrances of the hoistways are to be protected by substantial gates or bars erected to guard the full width of the landing entrance. All hoistway entrance bars and gates must be painted with diagonal contrasting colors, such as black and yellow stripes. In addition bars must not be less than two by four inch wooden bars, or the equivalent, located no less than two feet from the hoistway line. Locate bars not less than 36 inches, nor more than 42 inches above the floor. Gates or bars protecting the entrances to hoistways are to be equipped with latching devices.

An overhead protective covering of two-inch planking, 3/4-inch plywood, or other solid material of equivalent strength, shall be provided on the top of every material hoist cage or platform. The operator's station of a hoisting machine is to be provided with overhead protection equivalent to tight planking not less than two inches thick. Support for overhead protection is to be of equal strength.

Hoist towers may be used with or without an enclosure on all sides. However, the following conditions must be met:

1. When a hoist tower is enclosed, it is to be enclosed on all sides for its entire height, with a screen enclosure of 1/2-inch mesh, number 18 U.S. gauge wire, or equivalent, except for landing access.

2. When a hoist tower is not enclosed, the hoist platform or car shall be totally enclosed (caged) on all sides, for the full height between the floor and the overhead protective covering, with 1/2-inch mesh of number 14 U.S. gauge wire or equivalent. The hoist platform enclosure must include the required gates for loading and unloading. A six-foot high enclosure must be provided on the unused sides of the hoist tower at ground level.

Car arresting devices are to be installed to function in case of rope failure. All material hoist towers shall be designed by a licensed professional engineer. All material hoists must conform to the requirements of ANSI A10.5-1969, Safety Requirements for Material Hoists.

OVERHEAD HOIST (CFR 1926.554)

When operating an overhead hoist, the safe working load, as determined by the manufacturer, must be indicated on the hoist, and this safe working load shall not be exceeded. Also, the supporting structure to which the hoist is attached must have a safe working load equal to that of the hoist. The support must be arranged to provide free movement of the hoist, while not restricting it from aligning itself with the load. Overhead hoists are to be installed only in locations that permit the operator to stand clear of the load at all times.

An air-driven hoist must be connected to an air supply of sufficient capacity and pressure in order to safely operate the hoist. All air hoses are to be securely connected to prevent disconnection during use.

All overhead hoists in use must meet the applicable requirements for construction, design, installation, testing, inspection, maintenance, and operation, prescribed by their manufacturers. (See Figure 8-5 and Figure 8-6.)

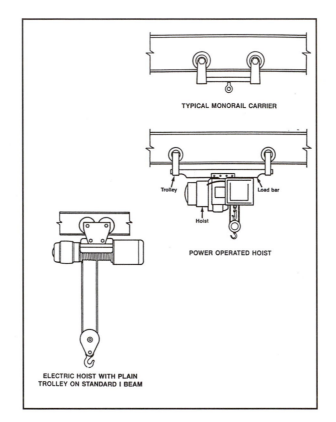

Figure 8-5. Overhead hoist. Courtesy of the Occupational Safety and Health Administration

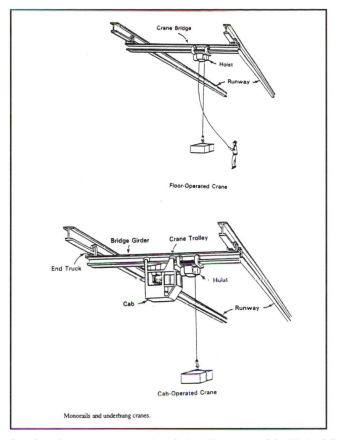

*Figure 8-6. Overhead and gantry cranes using hoist. Courtesy of the United States
        Department of Energy*

REFERENCES

Reese, C. D. and J.V. Eidson. *Handbook of OSHA Construction Safety & Health.*
Boca Raton, FL: CRC/Lewis Publishers, 1999.

United States Department of Energy. *Hoisting and Rigging Manual.* Washington, DC,
1991.

United States Department of Labor. Occupational Safety and Health Administration.
OSHA 10 and 30 Hour Construction Safety and Health Outreach Training Manual.
Washington, DC, 1991.

United States Department of Labor. Occupational Safety and Health Administration.
General Industry. *Code of Federal Regulations.* Title 29, Part 1910. Washington,
GPO, 1998.

United States Department of Labor. Occupational Safety and Health Administration.
Construction. *Code of Federal Regulations.* Title 29, Part 1926. Washington, GPO,
1998.

United States Department of Labor. Occupational Safety and Health Administration.
Office of Training and Education. *OSHA Voluntary Compliance Outreach Program:
Instructors Reference Manual.* Des Plaines, IL, 1993

CHAPTER 9

RIGGING

Ironworkers rigging a load

Using slings to rig equipment for lifting and handling requires special skill and knowledge. Whether made of wire rope, chain, or webbed material, a sling is a piece of equipment that needs a qualified (operator) person to use it, inspect it, and maintain it. OSHA's most common citations are given when no one has inspected riggings, when there is no competent person, when no tags appear on chains or synthetic slings, when defective alloy chains are used, when the company manufactures its own lifting devices, and when deaths due to overloading or defective rigging occur. Therefore it is important to plan and maintain lifting devices, inspect, and remove from service any faulty or damaged rigging.

The old saying, "rig to the center of gravity," is critical; the load must be level. Fouling of the load or rigging gear increases loading and can affect load control and safety. Wind, temperature, and dynamic loading conditions can also affect load distribution. Gravity and distribution are the essence of safe rigging processes.

Hardware is also an issue. When the load is placed in a sling, check connecting hardware, which can be dramatically affected by the sling angle and/or all multiple legs. Sling angles, between the legs and load, should not be less then 30 degrees. (See Figure 9-1 for calculating the stress on the legs of a sling.) Further information on load capacities and sling angles can be found in the section herein on chains, wire ropes, and web slings. All specifications for load limits are based on in-line loading. When side loading is allowed, the capacity must be reduced by at least 75 percent. Failure to consider hardware limitations can be costly.

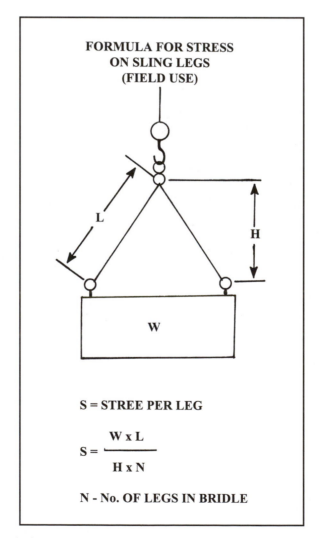

Figure 9-1. Formula for stress on sling legs (field use). Courtesy of the Occupational Safety and Health Administration

The ability to handle materials and to move them from one location to another, whether during transit or at the worksite, is vital to all segments of industry. For example, materials must be moved in order for industry to manufacture, sell, and utilize products. Without materials-handling capability, industry would cease to exist.

Mishandling of materials is the single largest cause of accidents and injuries in the workplace. All employees, regardless of workplace, take part in materials handling to varying degrees. As a result, some employees are injured. Most accidents and injuries, including associated physical pain, the loss of salary, and lost productivity, can be readily avoided. Whenever possible, mechanical means should be used to move materials and avoid employee injuries such as muscle pulls, strains, and sprains. In addition, many loads are too heavy and/or bulky to be safely moved manually. Therefore, various types of equipment have been designed to move specific types of materials. They include: cranes, derricks, hoists, powered industrial trucks, and conveyors.

Because cranes, derricks, and hoists rely upon slings to hold their suspended loads, slings are the most commonly used piece of materials-handling apparatuses. This chapter focuses on the proper selection, maintenance, and use of slings.

IMPORTANCE OF THE OPERATOR

The operator must exercise intelligence, care, and common sense in selecting and using slings. Select slings according to their intended use, the size and type of load, and the environmental conditions of the workplace. Visually inspect all slings before use to ensure that there is no obvious damage. Knowledge and experience precede all rigging processes.

A well-trained operator, for example, can prolong the service life of equipment and reduce costs by avoiding potentially hazardous effects caused by overloading equipment, operating it at excessive speeds, taking up slack with an abrupt sudden jerk, and the sudden accelerating or decelerating of equipment. The operator can anticipate causes and make adjustments to avoid danger. He or she should cooperate and communicate well with co-workers and supervisors and become a leader in carrying out safety measures, not merely for the good of the equipment and the production schedule, but, more importantly, for the safety of everyone concerned.

Rigging of loads shall be performed only by qualified riggers or qualified rigging specialists. The use of the word "rigger" is not intended to indicate a job classification; therefore, the heading and use here is a generic one. The title and training requirements apply to any employee who performs a rigging function. Qualified riggers shall meet the following requirements:

- Age—Be at least 18 years.
- Language—Understand spoken and written English or a language in use at the location.
- Knowledge—Have a basic knowledge and understanding of equipment-operational characteristics, capabilities, and limitations of the equipment. Understand rigging principles as applicable to the job for which they are to be qualified.
- Skill—Demonstrate skill in utilizing rigging principles.

Initial qualification of riggers shall include:

- Training with the equipment for the job for which they are to be qualified, under the direction of a qualified rigger or qualified rigging specialist designated by management.

- A review of the applicant's knowledge, including written and oral examinations, and observation of their skills by an instructor or examiner.

- A written record of training, competency, and authorization, which is inserted into the employee's training record by his or her supervisor. This record shall include identification of operations for which the individual is qualified.

Rigger qualification applies for a period of three years, unless it is revoked sooner by the rigger's manager. If a rigger is disqualified, his or her manager shall enter the action into the individual's personnel file. The program for maintenance of qualified-riggers status shall include:

- Verification by the rigger's manager that he or she is required to do rigging in the performance of duties, and that he or she has been doing so safely and competently.

- Participation by the qualified rigger in an approved training program.

Qualified Rigger Training Programs

All organizations employing personnel who do any rigging shall develop training programs, including means of testing, to assure that each rigger is competent to perform the operations. Safety personnel of the organization shall review program content for safety significance and include in their routine audits an evaluation of administrative policies and of each riggers ability to comply with the training and qualification program established and approved by the cognizant manager. The training program includes, but is not limited to:

- Proper usage of rope, shackles, hooks, wire rope, chain and fabric slings, timbers, hoisting principles, rollers, and scaffolds.

- How to calculate weight estimation, center of gravity, factors of safety, and the effect of angular pulls in load lifting.

- Safe attachment of slings for straight lifts, basket hitches, chokers, and multiple-bridle lifting.

- Safe and unsafe placement of sling hooks.

- Hook safety latches and hook mousing for safety.

- Load-sling adjustments to keep the center of gravity in line of hook pull.

- Risks and precautions when rigging near overhead power transmission lines.

- Use of spider bars for lifts and adjustment to keep the center of gravity in correct alignment for balanced lifts.

- Testing and inspecting rigging equipment in accordance with proper procedures.

SLING TYPES

The dominant characteristics of a sling are determined by the components of that sling. For example, the strengths and weaknesses of a wire rope sling are essentially the same as the strengths and weaknesses of the wire rope of which it is made. Slings are generally one of six types:

- Chain.

- Wire rope.

- Metal mesh.
- Natural fiber rope.
- Synthetic fiber rope.
- Synthetic web.

In general, use and inspection procedures tend to place these slings into three groups: chain, wire rope and mesh, and fiber rope web. Each type has its own particular advantages and disadvantages. Factors that should be taken into consideration when choosing the best sling for the job include the size, weight, shape, temperature, and sensitivity of the material to be moved, as well as the environmental conditions under which the sling will be used. (See Figure 9-2.)

Figure 9-2. Example of a sling in use

SAFE LIFTING PRACTICES

Once the sling is selected (based upon the characteristics of the load and the environmental conditions surrounding the lift) and inspected prior to use, the next step is learning how to use it safely. There are four primary factors to take into consideration for safely lifting a load. They are:

1. The size, weight, and center of gravity of the load.

2. The number of legs and the angle the sling makes with the horizontal line.

3. The rated capacity of the sling.

4. The history of the care and usage of the sling.

Size, Weight, and Center of Gravity of the Load

The center of gravity of an object is that point at which the entire weight may be considered as concentrated. In order to make a level lift, the crane hook must be directly above this point. While slight variations are usually permissible, if the crane hook is too far to one side of the center of gravity, dangerous tilting will result, causing unequal stresses in the different sling legs. This imbalance must be compensated for at once.

Number of Legs and Angle with the Horizontal

As the angle formed by the sling leg and the horizontal line decreases, the rated capacity of the sling also decreases. In other words, the smaller the angle between the sling leg and the horizontal, the greater the stress on the sling leg and the smaller (lighter) the load the sling can safely support. Larger (heavier) loads can be safely moved if the weight of the load is distributed among more sling legs.

Rated Capacity of the Sling

The rated capacity of a sling varies depending upon the type of sling, the size of the sling, and the type of hitch. Operators must know the capacity of the sling. Charts or tables that contain this information generally are available from sling manufacturers. The values given are for new slings. Older slings must be used with additional caution. Under no circumstances shall a sling's rated capacity be exceeded.

History of Care and Usage

The mishandling and misuse of slings is the leading cause of accidents involving their use. The majority of injuries and accidents, however, can be avoided by becoming familiar with the essentials of proper sling care and usage.

Proper care and usage are essential for maximum service and safety. Slings must be protected from sharp bends and cutting edges by means of cover saddles, burlap padding, or wood blocking, as well as from unsafe lifting procedures such as overloading.

Before making a lift, check to be certain that the sling is properly secured around the load and that the weight and balance of the load have been accurately determined. If the load is on the ground, do not allow the load to drag along the ground. This could damage the sling. If the load is already resting on the sling, be sure that there is no sling damage prior to making the lift.

Next, position the hook directly over the load and seat the sling squarely within the hook bowl. This gives the operator maximum lifting efficiency without bending the hook or overstressing the sling.

Wire rope slings are also subject to damage resulting from contact with sharp edges of the loads being lifted. These edges can be blocked or padded to minimize damage to the sling.

After the sling is properly attached to the load, there are a number of good lifting techniques that are common to all slings:

1.   Make sure that the load is not lagged, clamped, or bolted to the floor.

2.   Guard against shock loading by taking up the slack in the sling slowly.

3.   Apply power cautiously so as to prevent jerking at the beginning of the lift, and accelerate or decelerate slowly.

4.   Check the tension on the sling. Raise the load a few inches, stop, and check for proper balance and that all items are clear of the path of travel. Never allow anyone to ride on the hood or load.

5.   Keep all personnel clear while the load is being raised, moved, or lowered. Crane or hoist operators should watch the load at all times when it is in motion.

6.   Finally, obey the following "nevers:"

   a   Never allow more than one person to control a lift or give signals to a crane or hoist operator except to warn of a hazardous situation.

   b.   Never raise the load more than necessary.

   c.   Never leave the load suspended in the air.

   d.   Never work under a suspended load or allow anyone else to do so.

Once the lift has been completed, clean the sling, check it for damage, and store it in a clean, dry, airy place. It is best to hang it on a rack or wall. Remember, damaged slings cannot lift as much as new or well-cared-for older slings. Safe and proper use and storage of slings will increase their service life.

SAFE OPERATING PRACTICES FOR SLINGS

This section applies to slings used in conjunction with other material handling equipment for the movement of material by hoisting, in employments covered by this part. The types of slings covered are those made from alloy steel chain, wire rope, metal mesh, natural or synthetic fiber rope (conventional three strand construction), and synthetic web (nylon, polyester, and polypropylene). Whenever any sling is used, the following practices shall be observed:

1.   Slings that are damaged or defective shall not be used.

2.   Slings shall not be shortened with knots or bolts or other makeshift devices.

3.   Sling legs shall not be kinked.

4.   Slings shall not be loaded in excess of their rated capacities.

5.   Slings used in a basket hitch shall have the loads balanced to prevent slippage.

6.   Slings shall be padded or protected from the sharp edges of their loads.

7.   Slings shall be securely attached to their load.

8.   Suspended loads shall be kept clear of all obstructions.

9.   All employees shall be kept clear of loads about to be lifted and of suspended loads.

10.   Hands or fingers shall not be placed between the sling and its load while the sling is being tightened around the load.

11.   Shock loading is prohibited.

12.   A sling shall not be pulled from under a load when the load is resting on the sling.

Inspections

Each day before being used, the sling and all fastenings and attachments shall be inspected for damage or defects by a competent person designated by the employer. Additional inspections shall be performed during sling use, where service conditions warrant. Damaged or defective slings shall be immediately removed from service.

CHAIN SLINGS

Chains are commonly used because of their strength and ability to adapt to the shape of the load. Care should be taken, however, when using alloy chain slings because they are subject to damage by sudden shocks. Misuse of chain slings could damage the sling, resulting in sling failure and possible injury to an employee.

Chain slings are your best choice for lifting materials that are very hot. They can be heated to temperatures of up to 1,000 degrees Fahrenheit; however, when alloy chain slings are consistently exposed to service temperatures in excess of 600 degrees Fahrenheit, operators must reduce the working load limits in accordance with the manufacturer's recommendations.

Chains should have a safety factor of four to one. All alloy steel chains should have an identification tag, which is durable and permanently affixed and states the size, grade, rated capacity, and reach. Chains should never be used to lift loads which are not equal to or less than the manufacturer's rated capacities. Any hooks, rings, oblong links, pear shaped links, welded or mechanical coupling links, or other attachments are to be rated at least equal to the chain being used. If this is not the case, then the chains' rated lifting capacity must be reduced to the weakest component.

All sling types must be visually inspected prior to use. When inspecting alloy steel chain slings, pay special attention to any stretching, or wear in excess of the allowances made by the manufacturer. Also look for nicks and gouges. These are all indications that the sling may be unsafe and is to be removed from service.

Chain slings in particular must be cleaned prior to each inspection, as dirt or oil may hide damage. Then the operator must be certain to inspect the total length of the sling, periodically, looking for stretching, binding, wear, nicks, and gouges. If a chain sling has stretched so that it is now more than three percent longer than it was when new, it is unsafe and must be discarded.

Binding is the term used to describe the condition that exists when a sling has become deformed to the extent that its individual links cannot move within each other freely. It is also an indication that the sling is unsafe. Generally, wear occurs on the load-bearing inside ends of the links. Pushing links together so that the inside surface becomes clearly visible is the best way to check for this type of wear. Wear may also occur, however, on the outside of links when the chain is dragged along abrasive surfaces or pulled out from under heavy loads. Either type of wear weakens slings and makes accidents more likely.

Heavy nicks and gouges must be filed smooth. Measure the links with calipers, then compared the measurements with the manufacturer's minimum allowable safe dimensions. When in doubt, and in borderline situations, do not use the sling. In addition, never attempt to repair the welded components on a sling. If the sling needs repair of this nature, the supervisor must be notified.

Alloy steel chain slings with cracked or deformed master links, coupling link, or other components shall be removed from service. Worn or damaged alloy steel chain slings or attachments shall not be used until repaired.

Rated capacity (working load limit) for alloy steel chain slings must conform to the values shown in Figure 9-3. Whenever wear at any point of any chain link exceeds that shown in Figure 9-11 herein, from the *Crosby Lifting Guide*, and in compliance with Tables 184 -1 and 184 -2 in OSHA's 29 CFR 1910.184, the assembly is to be removed from service.

CHAIN SLING CAPACITIES (LBS.) – ANSI B30.9 DESIGN FACTOR 4/1

Crosby Q T ALLOY

4

CHAIN SIZE — CHAIN GR–8 DESIGN FACTOR 4/1	VERTICAL (SINGLE LEG)	TWO LEG OR BASKET HITCH (90°)	60 DEGREE SLING ANGLE	45 DEGREE SLING ANGLE	30 DEGREE S_LING ANGLE	SINGLE LEG MASTER LINK SIZE	DOUBLE LEG MASTER LINK SIZE
1/4 - (9/32)	3500	7000	6050	4900	3500	1/2	1/2
3/8	7100	14200	12200	10000	7100	3/4	3/4
1/2	12000	24000	20750	16950	12000	1	1
5/8	18100	39200	31350	25500	18100		1-1/4
3/4	28300	56600	49000	40000	28300	1-1/4	1-1/2
7/8	34200	68400	59200	48350	34200		1-3/4
1	47700	95400	82600	67450	47700	1-1/2	
1-1/4	72300	144600	125200	102200	72300		

TABLE 1
MAXIMUM ALLOWABLE WEAR AT ANY POINT OF LINK

NORMAL CHAIN OR COUPLING LINK CROSS SECTION	MAXIMUM ALLOWABLE WEAR DIAMETER INCHES
9/32	.037
3/8	.052
1/2	.069
5/8	.084
3/4	.105
7/8	.116
1	.137
1-1/4	.169

REFER TO ANSI B30.9 FOR FULL DETAILS
HORIZONTAL SLING ANGLES OF LESS THAN 30 DEGREES ARE NOT RECOMMENDED

CHAIN – FACTS

INSPECTION AND REMOVAL FROM SERVICE PER ANSI N30.9

FREQUENT INSPECTION

DAILY CHECK CHAIN AND ATTACHMENTS FOR WEAR, NICKS, CRACKS, BREAKS, GOUGES, STRETCH, BENDS, WELD SPLATTER, DISCOLORATION FROM EXCESSIVE TEMPERATURE, AND THROAT OPENINGS OF HOOKS.

1. CHAIN LINKS AND ATTACHMENTS SHOULD HINGE FREELY TO ADJACENT LINKS.
2. LATCHES ON HOOKS, IF PRESENT SHOULD HINGE FREELY AND SEAT PROPERLY WITHOUT EVIDENCE OF PERMANENT DISTORTION.

PERIODIC INSPECTION - INSPECTION RECORDS REQUIRED

• NORMAL SERVICE - YEARLY
• SEVERE SERVICE - MONTHLY

THIS INSPECTION SHALL INCLUDE EVERYTHING IN A FREQUENT INSPECTION PLUS EACH LINK AND END ATTACHMENT SHALL BE EXAMINED INDIVIDUALLY, TAKING CARE TO EXPOSE INNER LINK SURFACES OF THE CHAIN AND CHAIN ATTACHMENTS

1. WORN LINKS SHOULD NOT EXCEED VALUES GIVEN IN TABLE 1 OR RECOMMENDED BY THE MANUFACTURER
2. SHARP TRANSVERSE NICKS AND GOUGES SHOULD BE ROUNDED OUT BY GRINDING AND THE DEPTH OF THE GRINDING SHOULD NOT EXCEED VALUES IN TABLE 1
3. HOOKS SHOULD BE INSPECTED IN ACCORDANCE WITH ANSI B30.10
4. IF PRESENT, LATCHES ON HOOKS SHOULD SEAT PROPERLY, ROTATE FREELY, AND SHOW NO PERMANENT DISTORTION

Figure 9-3. Chain sling capacities. Permission by the Crosby Group, Inc.

In addition to a daily inspections, a thorough periodic inspection of alloy steel chain slings in use is to be made on a regular basis. The inspection is to be determined by the frequency of sling use, the severity of service conditions, the nature of the lifts being made, and the experience gained on the service life of slings used in similar circumstances. Such inspections are not to be performed at intervals no greater than 12 months apart. The employer must make and maintain a record of the most recent month in which each alloy steel chain sling is thoroughly inspected, and make such records available for examination by OSHA and other pertinent examiners.

WIRE ROPE SLINGS

Wire rope is composed of individual strands of wire which are composed of individual wires that are twisted to form strands. Strands are then twisted to form a wire rope. When wire rope has a fiber core, it is usually more flexible but is less resistant to environmental damage. Conversely, a core that is made of a wire rope strand tends to have greater strength and is more resistant to heat damage.

When wire rope is used, safe working loads must be determined for the various sizes and classifications of improved plow steel wire rope and wire rope slings with various types of terminals. For different sizes, classifications, and grades of wire rope, the safe working load recommended by the manufacturer for specific, identifiable products is to be followed, provided that a safety factor of not less than five is maintained.

Wire rope slings shall not be used with loads in excess of the rated capacities shown in Tables N-184-3 through N-184-14 of 29 CFR 1910.184. Slings not included in these tables shall be used only in accordance with the manufacturer's recommendations. Some general guidelines for wire ropes can be found in Figure 9-4.

Rope Lay

Wire rope may be further defined by the "lay." The lay of a wire rope can mean any of three things:

- One complete wrap of a strand around the core — One rope lay is one complete wrap of a strand around the core. (See Figure 9-5.)

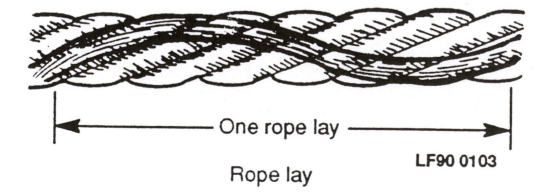

Figure 9-5. A rope lay. Courtesy of the Occupational Safety and Health Administration

WIRE ROPE SLING CAPACITIES (LBS.) – FLEMISH EYE – ANSI B30.9

6 X 19 AND 6 X 37
IMPROVED PLOW STEEL – IWRC 5/1 DESIGN FACTOR

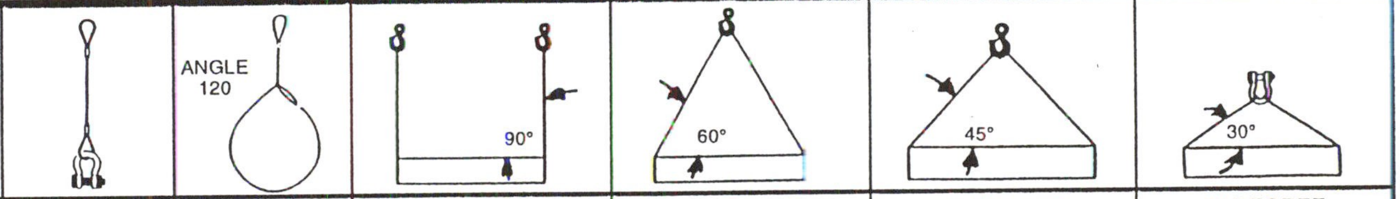

WIRE ROPE SIZE	Crosby Q & T CARBON SHACKLE MINIMUM SHACKLE SIZE FOR A D/d > 1 AT LOAD CONNECTION — SHACKLE SIZE	VERTICAL (SINGLE LEG)	CHOKER (ANGLE 120)	TWO LEG OR BASKET HITCH (90°)	60 DEGREE SLING ANGLE	45 DEGREE SLING ANGLE	30 DEGREE SLING ANGLE
1/4	5/16	1120	820	2200	1940	1500	1100
5/16	3/8	1740	1280	3400	3000	2400	1700
3/8	7/16	2400	1840	4800	4200	3400	2400
7/16	1/2	3400	2400	6800	5800	4800	3400
1/2	5/8	4400	3200	8800	7600	6200	4400
9/16	5/8	5600	4000	11200	9600	7900	5600
5/8	3/4	6800	5000	13600	11800	9600	6800
3/4	7/8	9800	7200	19600	16900	13800	9800
7/8	1	13200	9600	26400	22800	18600	13200
1	1-1/8	17000	12600	34000	30000	24000	17000
1-1/8	1-1/4	20000	15800	40000	34600	28300	20000
1-1/4	1-3/8	26000	19400	52000	45000	36700	26000
1-3/8	1-1/2	30000	24000	60000	52000	42400	30000

- RATED CAPACITIES BASED ON PIN DIAMETER OR HOOK NO LARGER THAN THE NATURAL EYE WIDTH (1/2 X EYE LENGTH) OR LESS THAN THE NOMINAL SLING DIAMETER

REFER TO ANSI B30.9 FOR FULL DETAILS

HORIZONTAL SLING ANGLES OF LESS THAN 30 DEGREES ARE NOT RECOMMENDED

Figure 9-4. Wire rope capacities. Permission by the Crosby Group, Inc.

- The direction in which the strands are wound around the core—Wire rope is referred to as right lay or left lay. A right lay rope is one in which the strands are wound in a right-hand direction like a conventional screw thread. A left lay rope is just the opposite.

- The direction the wires are wound in the strands in relation to the direction of the strands around the core—In regular lay rope, the wires in the strands are laid in one direction while the strands in the rope are laid in the opposite direction. In lang lay rope, the wires are twisted in the same direction as the strands.

In regular lay ropes, where the wires in the strands are laid in one direction, and strands in the rope are laid in the opposite direction, the end result is a wire crown running approximately parallel to the longitudinal axis of the rope. These ropes have good resistance to kinking and twisting and are easy to handle. They are also able to withstand considerable crushing and distortion due to the short length of exposed wires. Consequently, this type of rope has the widest range of applications.

Lang lay ropes, where the wires are twisted in the same direction as the strands, is recommended for many excavating, construction, and mining applications, including draglines, hoist lines, dredgelines, and other similar lines.

Lang lay ropes are more flexible and have a greater wearing surface per wire than regular lay ropes. In addition, since the outside wires in lang lay rope lie at an angle to the rope axis, internal stress due to bending over sheaves and drums is reduced, causing lang lay ropes to be more resistant to bending fatigue.

A left lay rope is one in which the strands form a left-hand helix similar to the threads of left-hand screw threads. Left lay rope has its greatest usage in oil fields on rod and tubing lines, blast hole rigs, and on spudders where rotation of right lay would loosen couplings. Rotation of a left lay rope tightens a standard coupling. (See Figure 9-6.)

Safe Operating Temperatures

Fiber core wire rope slings of all grades shall be permanently removed from service if they are exposed to temperatures in excess of 200 degrees Fahrenheit. When nonfiber core wire rope slings of any grade are used at temperatures above 400 degrees Fahrenheit, or below minus 60 degrees Fahrenheit, recommendations of the sling manufacturer regarding use at these temperature shall be followed.

Inspections and Maintenance

Wire rope slings, like chain slings, must be cleaned prior to each inspection because they are also subject to damage hidden by dirt or oil. In addition, they must be lubricated according to manufacturer's instructions. Lubrication prevents or reduces corrosion and wear due to friction and abrasion. Before applying any lubricant, however, the sling user should make certain that the sling is dry. Applying lubricant to a wet or damp sling traps moisture against the metal and hastens corrosion.

Corrosion deteriorates wire rope. It may be indicated by pitting, but it is sometimes hard to detect. Therefore, if a wire rope sling shows any sign of significant deterioration, it must be removed until it can be examined by a qualified person to determine the extent of damage.

Wire rope slings must be visually inspected before each use. The operator should check the twists or lay of the sling. If ten randomly distributed wires in one lay are broken, or five wires in one strand of a rope lay are damaged, the sling must not be used. It is not suffi-

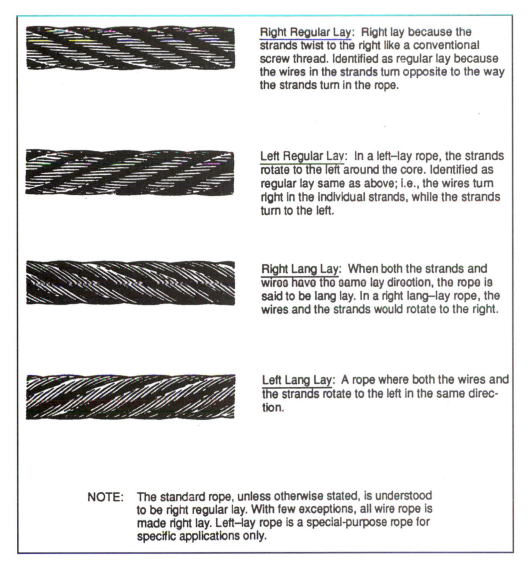

Right Regular Lay: Right lay because the strands twist to the right like a conventional screw thread. Identified as regular lay because the wires in the strands turn opposite to the way the strands turn in the rope.

Left Regular Lay: In a left–lay rope, the strands rotate to the left around the core. Identified as regular lay same as above; i.e., the wires turn right in the individual strands, while the strands turn to the left.

Right Lang Lay: When both the strands and wires have the same lay direction, the rope is said to be lang lay. In a right lang–lay rope, the wires and the strands would rotate to the right.

Left Lang Lay: A rope where both the wires and the strands rotate to the left in the same direction.

NOTE: The standard rope, unless otherwise stated, is understood to be right regular lay. With few exceptions, all wire rope is made right lay. Left–lay rope is a special-purpose rope for specific applications only.

Figure 9-6. Examples of types of rope lays. Courtesy of the Occupational Safety and Health Administration

cient, however, to check only the condition of the wire rope. End fittings and other components should also be inspected for any damage that could make the sling unsafe.

To ensure safe sling usage between scheduled inspections, all workers must participate in a safety awareness program and report potential problems. Each operator must keep a close watch on slings he or she is using. If any accident involving the movement of materials occurs, the operator must immediately shut down the equipment and report the accident to a supervisor. The cause of the accident must be determined and corrected before resuming operations. By applying these guidelines to proper sling use and maintenance, and by the avoidance of kinking, it is possible to greatly extend a wire rope sling's useful service life.

Wire Rope Life

Many operating conditions affect the life of wire rope, such as bending, stresses, loading conditions, speed of load application (jerking), abrasion, corrosion, sling design, materials handled, environmental conditions, and history of previous usage. The weight, size, and shape of loads handled affect the service life of a wire rope sling, and flexibility is also a factor. Generally, more flexible ropes are selected when smaller radius bending is required, and less flexible ropes should be used when the rope must move through or over abrasive materials.

Removal from Service

Wire rope slings shall be immediately removed from service if any of the following conditions are present:

- Ten randomly distributed broken wires in one rope lay, or five broken wires in one strand in one rope lay.

- Wear or scraping of 1/3 the original diameter of outside individual wires.

- Kinking, crushing, birdcaging or any other damage.

- Evidence of heat damage.

- End attachments that are cracked, deformed, or worn.

- Hooks that have been opened more than 15 percent of the normal throat opening measured at the narrowest point or twisted more than ten degrees from the plane of the unbent hook.

- Corrosion of the rope or end attachments.

Wire Rope Sling Selection

When selecting a wire rope sling to give the best service, there are four characteristics to consider: strength, ability to bend without distortion, ability to withstand abrasive wear, and ability to withstand abuse.

Strength

The strength of a wire rope is a function of its size, grade, and construction. It must be sufficient to accommodate the maximum load that will be applied. The maximum load limit is determined by means of an appropriate multiplier. This multiplier is the number by which the ultimate strength of a wire rope is divided to determine the working load limit. Thus a wire rope sling with a strength of 10,000 pounds and a total working load of 2,000 pounds has a design factor (multiplier) of five. New wire rope slings have a design factor of five. As a sling suffers from the rigors of continued service, however, both the design factor and the sling's ultimate strength are proportionately reduced. If a sling is loaded beyond its ultimate strength, it will fail. For this reason, older slings must be more rigorously inspected to ensure that rope conditions adversely affecting the strength of the sling are considered in determining whether or not a wire rope sling should be allowed to continue in service.

Fatigue

A wire rope must have the ability to withstand repeated bending without the failure of its wires from fatigue. Fatigue failure of wires results after the development of small cracks under repeated applications of bending loads, when a rope makes small radius bends. The best means of preventing fatigue failure of wire rope slings is to use blocking or padding to increase the radius of the bend.

<u>Abrasive Wear</u>

The ability of a wire rope to withstand abrasion is determined by the size, number of wires, and construction of the rope. Smaller wires bend more readily and therefore offer greater flexibility but are less able to withstand abrasive wear. Conversely, the larger wires of less flexible ropes are better able to withstand abrasion than smaller wires of more flexible ropes.

<u>Abuse</u>

All other factors being equal, misuse or abuse of wire rope will cause a wire rope sling to become unsafe long before any other factor. Abusing a wire rope sling can cause serious structural damage such as kinking or birdcaging which reduces the strength of the wire rope. (In birdcaging, the wire rope strands are forcibly untwisted and become spread outwards.) Therefore, in order to prolong the life of the sling and protect the lives of employees, the manufacturer's suggestions for safe and proper use of wire rope slings must be strictly adhered to.

Field Lubrication

Although every rope sling is lubricated during manufacture, to lengthen its useful service life it must also be lubricated "in the field." There is no set rule on how much or how often this should be done. It depends on the conditions under which the sling is used. The heavier the loads, the greater the number of bends, or the more adverse the conditions under which the sling operates, the more frequently lubrication will be required.

Storage

Wire rope slings should be stored in a well ventilated, dry building or shed. Never store them on the ground or allow them to be continuously exposed to the elements because this will make them vulnerable to corrosion and rust. If it is necessary to store wire rope slings outside, make sure that they are set off the ground and protected. Note: Using the sling several times a week, even at a light load, is a good practice. Records show that slings used frequently and continuously give useful service far longer than idle slings.

General Information

Slings must never be shortened with knots or bolts or other makeshift devices, nor are sling legs to be kinked. Slings used in a basket hitch must have their loads balanced to prevent slippage. They must be padded or protected from the sharp edges of their loads. Hands and fingers shall not be placed between the sling and its load while the sling is being tightened around the load. Shock loading is prohibited. Slings are not to be pulled from under a load when the load is resting on the sling.

Welding of end attachments, except covers to thimbles, is to be performed prior to the assembly of the sling. All welded end attachments are not to be used unless proof tested by the manufacturer, or an equivalent entity, at twice their rated capacity, prior to initial use. The employer must retain a certificate of proof of test, and make it available for examination by OSHA and other pertinent authorities.

Protruding ends of strands in splices on slings and bridles are to be covered or blunted. Wire rope is not to be secured by knots, except on haulback lines on scrapers. An eye splice made in any wire rope is not to be less than three full tucks. However, this requirement does not operate to preclude the use of another form of splice or connection which can be shown to be as efficient and which is not otherwise prohibited.

Except for eye splices at the ends of wires, and for endless rope slings, each wire rope used in hoisting, lowering, or pulling loads, must consist of one continuous piece without a knot or splice. Eyes in wire rope bridles, slings, or bull wires are not to be formed by wire rope clips or knots.

U Bolts

When U bolt wire rope clips are used to form eyes, information in Figure 9-7 can be used to determine the number and spacing of the clips based upon the wire's diameter. When used for eye splices, the U bolt is to be applied so that the "U" section is in contact with the dead end of the rope.

Metal Mesh Slings

Sling Marking

Each metal mesh sling shall have permanently affixed to it a durable marking that states the rated capacity for vertical basket hitch and choker hitch loadings.

Handles

Handles shall have a rated capacity at least equal to the metal fabric and exhibit no deformation after proof testing.

Attachments of Handles to Fabric

The fabric and handles shall be joined so that:
- The rated capacity of the sling is not reduced.
- The load is evenly distributed across the width of the fabric.
- Sharp edges will not damage the fabric.

Sling Coatings

Coatings which diminish the rated capacity of a sling shall not be applied.

Sling Testing

All new and repaired metal mesh slings, including handles, shall not be used unless proof tested by the manufacturer, or equivalent entity, at a minimum of 1 1/2 times their rated capacity. Elastomer impregnated slings shall be proof tested before coating.

Sling Use

Metal mesh slings shall not be used to lift loads in excess of their rated capacities as

Crosby® RIGGING HARDWARE

10

Crosby® WIRE ROPE CLIPS

SIZE	EFFICIENCY	NUMBER OF CLIPS	TURNBACK LENGTH (IN)	TORQUE FT – lbs
1/8	80%	2	3-1/4	4.5
3/16	80%	2	3-3/4	7.5
1/4	80%	2	4-3/4	15
5/16	80%	2	5-1/4	30
3/8	80%	2	6-1/2	45
7/16	80%	2	7	65
1/2	80%	3	11-1/2	65
9/16	80%	3	12	95
5/8	80%	3	12	95
3/4	80%	4	18	130
1	90%	5	26	225

Crosby® TURNBUCKLE

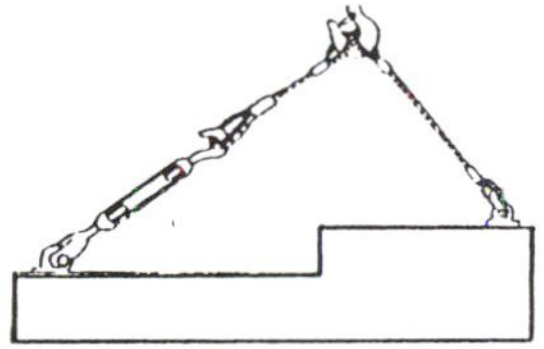

SIZE	WORKING LOAD LIMIT JAW AND EYE 5/1 DESIGN FACTOR	WORKING LOAD LIMIT HOOK END FITTING 5/1 DESIGN FACTOR
1/4	500	400
5/16	800	700
3/8	1200	1000
1/2	2200	1500
5/8	3500	2250
3/4	5200	3000
7/8	7200	4000
1	10000	5000
1-1/4	15200	5000
1-1/2	21400	7500

APPLY U-BOLT OVER DEAD END OF THE WIRE ROPE LIVE END OF THE ROPE RESTS IN THE SADDLE A TERMINATION IS NOT COMPLETE UNTIL IT HAS BEEN RETORQUED A SECOND TIME **NEVER SADDLE A DEAD HORSE!**

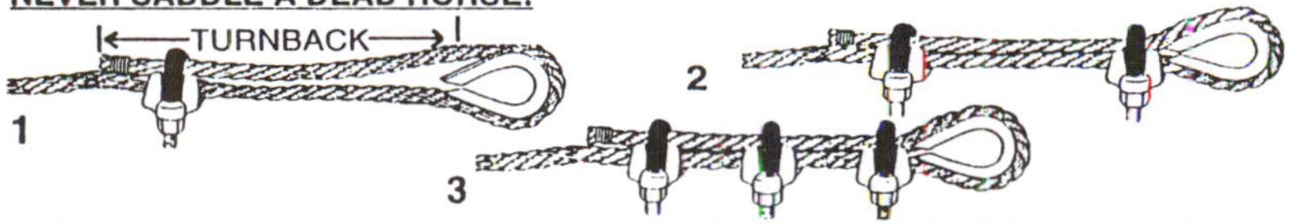

THE USE OF LOCKNUTS OR MOUSING IS AN EFFECTIVE METHOD OF PREVENTING TURNBUCKLES FROM ROTATING

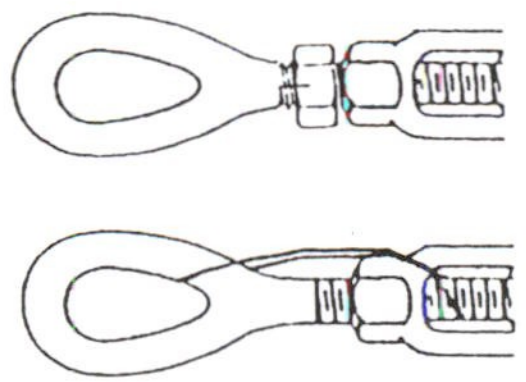

FOR ADDITIONAL INFORMATION REFER TO THE Crosby® PRODUCT WARNING

REFER TO THE Crosby® CATALOG FOR ADDITIONAL INFORMATION

Figure 9-7. Rigging hardware: wire rope clips. Permission by the Crosby Group, Inc.

prescribed in Table N-184-15 of 29 CFR 1910.184. Slings not included in this table shall be used only in accordance with the manufacturer's recommendations.

Safe Operating Temperatures

Metal mesh slings which are not impregnated with elastomers, may be used in a temperature range from minus 20 degrees Fahrenheit to plus 550 degrees Fahrenheit without decreasing the working load limit. Metal mesh slings impregnated with polyvinyl chloride or neoprene may be used only in a temperature range from zero degrees to plus 200 degrees Fahrenheit. For operations outside these temperature ranges, or for metal mesh slings impregnated with other materials, the sling manufacturer's recommendations shall be followed.

Repairs

Metal mesh slings which are repaired shall not be used unless repaired by a metal mesh sling manufacturer or an equivalent entity. Once repaired, records shall be maintained to indicate the date and nature of repairs and the person or organization who performed the repairs.

Removal from Service

Metal mesh slings shall be immediately removed from service if any of the following conditions are present:

- A broken weld or brazed joint along the sling edge.
- Reduction in wire diameter of 25 percent due to abrasion or 15 percent due to corrosion.
- Lack of flexibility due to distortion of the fabric.
- Distortion of the female handle so that the depth of the slot is increased more than ten percent.
- Distortion of either handle so that the width of the eye is decreased more than ten percent.
- A 15 percent reduction of the original cross sectional area of metal at any point around the handle eye.
- Distortion of either handle out of its plane.

FIBER ROPE AND SYNTHETIC WEB SLINGS

Fiber rope and synthetic web slings are used primarily for temporary work, such as construction and painting jobs, and in marine operations. They are also the best choice for use on expensive loads, highly finished parts, fragile parts, and delicate equipment.

Natural and Synthetic Fiber Rope Slings

Fiber rope slings are preferred for some applications because they are pliant, grip the load well, and do not mar the surface of the load. However they should be used only on light loads and must not be used on objects that have sharp edges capable of cutting the rope or in applications where the sling will be exposed to high temperatures, severe abrasion, or acids.

The choice of rope type and size depends upon the application, the weight to be lifted, and sling angle. Before lifting any load with a fiber rope sling, be sure to inspect the sling carefully because they deteriorate far more rapidly than wire rope slings, and their actual strength is very difficult to estimate.

Fiber rope slings made from conventional three strand construction fiber rope shall not be used with loads in excess of the rated capacities prescribed in Tables N-184-16 through N-184-19 of OSHA's 29 CFR 1910.184. Fiber rope slings shall have a diameter of curvature meeting at least the minimums specified in Figures N-184-4 and N-184-5 of OSHA's 29 CFR 1910.184. Slings not included in these tables shall be used only in accordance with the manufacturer's recommendations.

Safe Operating Temperatures

Natural and synthetic fiber rope slings, except for wet frozen slings, may be used in a temperature range from minus 20 degrees Fahrenheit to plus 180 degrees Fahrenheit without decreasing the working load limit. For operations outside this temperature range and for wet frozen slings, the sling manufacturer's recommendations shall be followed.

Splicing

Spliced fiber rope slings shall not be used unless it has been spliced in accordance with minimum requirements specified in OSHA's 29 CFR 1910.184 and with any additional recommendations of the manufacturer.

All splice made to rope slings, which are provided by the employer, are to be made in accordance with the manufacturer's fiber rope recommendations. In manila rope, eye splices are to contain at least three full tucks, and short splices must contain at least six full tucks (three on each side of the center line of the splice). In layed synthetic fiber rope, eye splices must contain at least four full tucks, and short splices must contain at least eight full tucks (four on each side of the center line of the splice). Strand end tails are not to be trimmed short (flush with the surface of the rope) but trimmed immediately adjacent to the full tucks. This precaution applies to both eye and short splices and all types of fiber rope. For fiber ropes under one-inch diameter, the tails must project at least six rope diameters beyond the last full tuck. For fiber ropes one-inch diameter and larger, the tails must project at least six inches beyond the last full tuck. In applications where the projecting tails may be objectionable, the tails are to be tapered and spliced into the body of the rope using at least two additional tucks (this will require a tail length of approximately six rope diameters beyond the last full tuck). For all eye splices, the eye is to be sufficiently large enough to provide an included angle of not greater than 60 degrees at the splice, when the eye is placed over the load or support. Knots are never to be used in lieu of splices.

Inspection

When inspecting a fiber rope sling prior to using it, look first at its surface. Look for dry, brittle, scorched, or discolored fibers. If any of these conditions are found, the supervisor must be notified and a determination made regarding the safety of the sling. If the sling is found to be unsafe, it must be discarded.

Next, check the interior of the sling. It should be as clean as when the rope was new. A build-up of powder-like sawdust on the inside of the fiber rope indicates excessive internal wear and is an indication that the sling is unsafe.

Finally, scratch the fibers with a fingernail. If the fibers come apart easily, the fiber sling has suffered some kind of chemical damage and must be discarded.

Removal from Service

Natural and synthetic fiber rope slings shall be immediately removed from service if any of the following conditions are present:

- Abnormal wear.

- Powdered fiber between strands.

- Variations in the size or roundness of strands.

- Discoloration or rotting.

- Distortion of hardware in the sling.

Repairs

Only fiber rope slings made from new rope shall be used. Use of repaired or reconditioned fiber rope slings is prohibited. Fiber ropes and synthetic webs are generally discarded rather than serviced or repaired. Operators must always follow manufacturer's recommendations.

Synthetic web slings illustrated in Figure N-184-6 shall not be used with loads in excess of the rated capacities specified in Tables N-184-20 through N-184-22 of OSHA's 29 CFR 1910.184. Slings not included in these tables shall be used only in accordance with the manufacturer's recommendations.

Synthetic Web Slings

Synthetic web slings offer a number of advantages for rigging purposes. The most commonly used synthetic web slings are made of nylon, dacron, and polyester. They have the following properties in common:

- Strength—can handle load of up to 300,000 pounds.

- Convenience—can conform to any shape.

- Safety—will adjust to the load contour and hold it with a tight, non-slip grip.

- Load protection—will not mar, deface, or scratch highly polished or delicate surfaces.

- Long life—are unaffected by mildew, rot, or bacteria; resist some chemical action; and have excellent abrasion resistance.

- Economy—have low initial cost plus long service life.

- Shock absorbency—can absorb heavy shocks without damage.

- Temperature resistance—are unaffected by temperatures up to 180 degrees Fahrenheit.

Web Sling Properties

Each synthetic material has its own unique properties. Nylon must be used wherever alkaline or greasy conditions exist. Nylon is also preferable when neutral conditions prevail and when resistance to chemicals and solvents is important. Dacron must be used where high concentrations of acid solutions—such as sulfuric, hydrochloric, nitric, and formic acids— and where high-temperature bleach solutions are prevalent. (Nylon will deteriorate under these conditions.) Do not use dacron in alkaline conditions because it will deteriorate; use nylon or polypropylene instead. Polyester must be used where acids or bleaching agents are present and is also ideal for applications where a minimum of stretching is important.

Web slings made from synthetic materials with a safety factor of five to one (ANSI B30.9A-1994) must be proof tested to twice the rated load. Rated capacities are not to be exceeded. Synthetic web slings are to be permanently marked with the manufacturer's name, stock number, type of material, and rate loads for different types of hitches. As with any sling choker, hitch loads are not to exceed 80 percent of vertical rated load. The rated loads for nonvertical bridle or basket hitches must be adjusted in accordance with the horizontal sling angle. The horizontal angle should never be less than 30 degrees. (See Figure 9-8.)

Synthetic webbing shall be of uniform thickness and width, and selvage edges must not be split from the webbing's width. Fittings shall be of a minimum breaking strength equal to that of the sling and must be free of all sharp edges that could in any way damage the webbing. Stitching is the only method permitted for attaching end fittings to webbing and to form eyes. The thread must be an even pattern and contain a sufficient number of stitches to develop the full breaking strength of the sling.

Web slings should be inspected when purchased, prior to each use, and annually. Although OSHA does not require recordkeeping, it is highly recommended. Inspections by a competent person(s) should be done according to the frequency of use, adverse lifting conditions, and other extenuating circumstances, such as temperature extremes. Synthetic web slings of polyester and nylon are not to be used at temperatures in excess of 180 degrees Fahrenheit. Polypropylene web slings are not to be used at temperatures in excess of 200 degrees Fahrenheit.

Possible Defects

The reliability of synthetic web slings can be permanently affected by any of the following defects:

1. Acid or caustic burns.

2. Melting or charring of any part of the surface.

3. Snags, punctures, tears, or cuts.

4. Broken or worn stitches.

5. Wear or elongation exceeding the amount recommended by the manufacturer.

6. Distortion of fittings.

Sling Identification

Each sling shall be marked or coded to show the rated capacities for each type of hitch and type of synthetic web material.

Environmental Conditions

When synthetic web slings are used, the following precautions shall be taken:

1. Nylon web slings shall not be used where fumes, vapors, sprays, mists or liquids of acids or Phenolics are present.

2. Polyester and polypropylene web slings shall not be used where fumes, vapors, sprays, mists or liquids of caustic chemicals are present.

3. Web slings with aluminum fittings shall not be used where fumes, vapors, sprays, mists or liquids of caustic chemicals are present.

WEB SLING CAPACITIES – ANSI B30.9 – DESIGN FACTOR 5/1

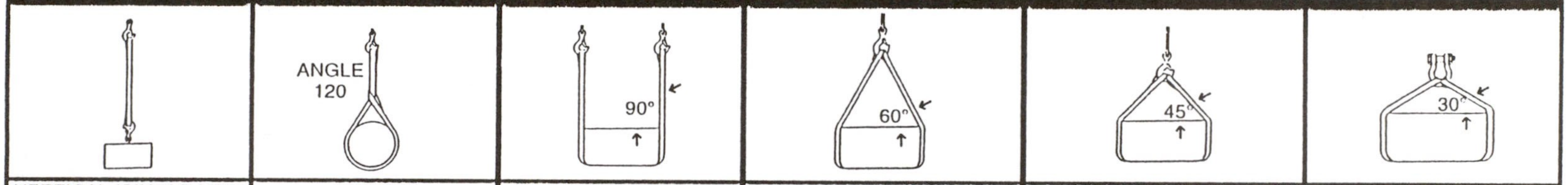

VERTICAL (SINGLE LEG)	CHOKER	TWO LEG OR BASKET	60 DEGREE SLING ANGLE	45 DEGREE SLING ANGLE	30 DEGREE SLING ANGLE
100% OF SINGLE LEG	80% OF SINGLE LEG	200% OF SINGLE LEG	170% OF SINGLE LEG	140% OF SINGLE LEG	SAME AS SINGLE LEG

INSPECTION AND REMOVAL FROM SERVICE PER ANSI B30.9

WEB SLING

FREQUENT INSPECTION
THIS INSPECTION SHALL BE MADE BY THE PERSON HANDLING THE SLING EACH DAY THE SLING IS USED

PERIODIC INSPECTION WRITTEN INSPECTION RECORDS SHOULD BE KEPT FOR ALL SLINGS
THIS INSPECTION SHOULD BE CONDUCTED BY DESIGNATED PERSONNEL, FREQUENCY OF THE INSPECTION SHOULD BE BASED THE FOLLOWING:
1. FREQUENCY OF SLING USE
2. SEVERITY OF SERVICE CONDITIONS
3. EXPERIENCE GAINED ON THE SERVICE LIFE OF SLING USED IN SIMILAR APPLICATIONS
4. AT LEAST ANNUALLY

REMOVAL CRITERIA
1. ACID OR CAUSTIC BURNS
2. MELTING OR CHARRING OF ANY PART OF THE SLING
3. BROKEN, TEARS, CUTS, OR SNAGS
4. BROKEN OR WORN STITCHING IN LOAD BEARING SPLICES
5. EXCESSIVE ABRASIVE WEAR
6. KNOTS IN ANY PART OF THE SLING
7. EXCESSIVE PITTING OR CORROSION, OR CRACKED DISTORTED OR BROKEN FITTINGS
8. OTHER VISIBLE DAMAGE THAT CAUSES DOUBT AS TO THE STRENGTH OF THE SLING

• DO NOT "BUNCH", OR "PINCH" THE SLING IN FITTINGS
• DO NOT PLACE EYE OVER A PIN OR HOOK GREATER THAN 1/2 TIMES EYE LENGTH

REFER TO ANSI B30.9 FOR FULL DETAILS

HORIZONTAL SLING ANGLES OF LESS THAN 30 DEGREES ARE NOT RECOMMENDED

Figure 9-8. Web sling capacities. Permission by the Crosby Group, Inc.

Safe Operating Temperatures

Synthetic web slings of polyester and nylon shall not be used at temperatures in excess of 180 degrees Fahrenheit. Polypropylene web slings shall not be used at temperatures in excess of 200 degrees Fahrenheit.

Inspection

When inspecting web slings, look for knots, stretching, broken or worn stitches, chemical burns, cuts, snags, holes, and tears or excessive abrasions. Inspection should include an examination of fittings for cracks, deforming, corrosion, pitting, breakage, and missing or illegible labels. If any of these are observed, the web sling should be removed from service immediately and destroyed or tagged out. Remember that ultraviolet or extreme sunlight can also damage web slings; this necessitates storing them out of sunlight and in a dry area. In all cases, the manufacturer's guidelines, rated loads, and specifications should be followed.

Repairs

Synthetic web slings that are repaired shall not be used unless repaired by the sling manufacturer or an equivalent entity.

Removal from Service

Synthetic web slings shall be immediately removed from service if any of the following conditions are present:

- Acid or caustic burns.
- Melting or charring of any part of the sling surface.
- Snags, punctures, tears, or cuts.
- Broken or worn stitches.
- Wear or elongation that exceeds the manufacturer's recommended limits.
- Distortion of fittings.

Shackles and Hooks (1926.251)

Figure 9-9 provides information regarding the safe working loads of various sizes of shackles. Higher safe working loads are permissible, when recommended by the manufacturer for specific, identifiable products (provided that a safety factor of not less than five is maintained) and hooks.

Manufacturers' recommendations are to be followed in determining what is a safe working load for the various sizes and types of specific and identifiable hooks. All hooks for which no applicable manufacturer's recommendations are available, are to be tested to twice the intended safe working load, before they are initially put into use. The employer shall maintain a record of the dates and results of such tests.

SUMMARY

Good safety practices protect workers and incidental passers by while using slings to move materials. First, learn as much as possible about the materials with which will be handled.

Crosby® RIGGING HARDWARE

Crosby® SHACKLES

SCREW PIN AND BOLT TYPE · QUENCHED AND TEMPERED · QUIC-CHECK®

NOMINAL SIZE (IN) DIAMETER OF BOWS	CARBON MAXIMUM WORKING LOAD TONS (CARBON SHACKLE DESIGN FACTOR 6/1)	ALLOY MAXIMUM WORKING LOAD TONS (ALLOY SHACKLE DESIGN FACTOR 5/1)	INSIDE WIDTH AT PIN (INCHES)	DIAMETER OF PIN
3/16	1/3		.38	.25
1/4	1/2		.47	.31
5/16	3/4		.53	.38
3/8	1	2	.66	.44
7/16	1-1/2	2.6	.75	.50
1/2	2	3.3	.81	.63
5/8	3-1/4	5	1.06	.75
3/4	4-3/4	7	1.25	.88
7/8	6-1/2	9.5	1.44	1.00
1	8-1/2	12.5	1.69	1.13
1-1/8	9-1/2	15	1.81	1.25
1-1/4	12	18	2.03	1.38
1-3/8	13-1/2	21	2.25	1.50
1-1/2	17	30	2.38	1.63

· INSURE SCREW PIN TIGHT BEFORE EACH LIFT
· USE BOLT TYPE SHACKLE FOR PERMANENT INSTALLATION 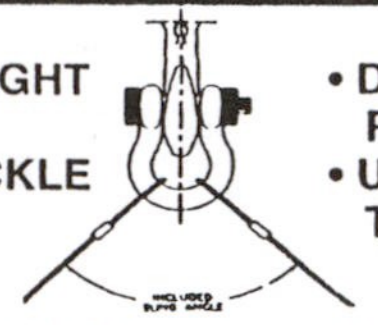
· DO NOT SIDE LOAD ROUND PIN SHACKLE
· USE SCREW PIN OR BOLT TYPE TO COLLECT SLINGS

MAXIMUM INCLUDED ANGLE 120 DEGREES

REFER TO Crosby® CATALOG FOR ADDITIONAL INFORMATION

Crosby® HOOKS

SHANK HOOK, SWIVEL HOOK, EYE HOOK · QUENCHED AND TEMPERED · QUIC-CHECK®

DESIGN FACTOR
· EYEHOOKS – 5/1 (EXCEPT ALLOY 30 TON AND LARGER ARE 4-1/2 /1
· SHANK AND SWIVEL ARE 4-1/2 /1

CARBON MAXIMUM WORKING LOAD TONS	CODE	ALLOY MAXIMUM WORKING LOAD TONS	CODE	THROAT OPENING (INCHES)	DEFORMATION INDICATOR A - A
3/4	DC	1	DA	.88	1.50
1	FC	1-1/2	FA	.97	2.00
1-1/2	GC	2	GA	1.00	2.00
2	HC	3	HA	1.12	2.00
3	IC	*4-1/2 /5	IA	1.06	2.50
5	JC	7	JA	1.50	3.00
7-1/2	KC	11	KA	1.75	4.00
10	LC	15	LA	1.91	4.00
15	NC	22	NA	2.75	5.00
20	OC	30	OA	3.25	6.50
25	PC	37	PA	3.00	7.00
30	SC	45	SA	3.38	8.00
40	TC	60	TA	4.12	10.00

* 320N EYE HOOK IS NOW RATED AT 5 TONS

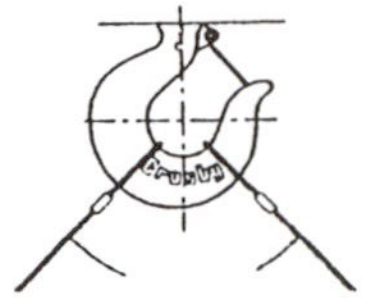

MAXIMUM INCLUDED ANGLE 90 DEGREES

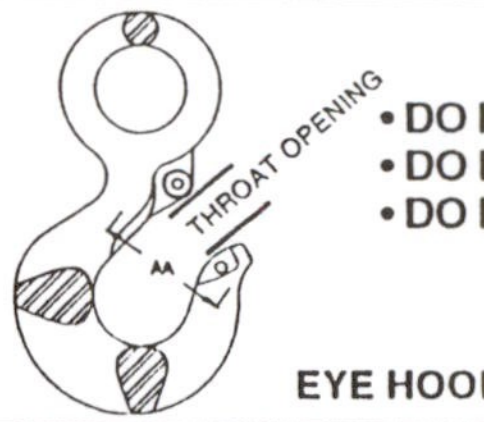

THROAT OPENING

· DO NOT SIDELOAD
· DO NOT TIP LOAD
· DO NOT BACKLOAD

EYE HOOK

REFER TO Crosby® GROUP PRODUCT WARNING FOR ADDITIONAL INFORMATION

Figure 9-9. Shackles and hooks. Permission by the Crosby Group, Inc.

Slings come in many different types, one of which is right for the job. Second, analyze the load to be moved in terms of size, weight, shape, temperature, and sensitivity, then choose the sling which best meets those needs. Third, always inspect all the equipment before and after a move. Always be sure to give equipment whatever in-service maintenance it may need. Fourth, use safe lifting practices. Use the proper lifting technique for the type of sling and the type of load.

REFERENCES

Reese, C. D. and J.V. Eidson. *Handbook of OSHA Construction Safety & Health*. Boca Raton, FL: CRC/Lewis Publishers, 1999.

The Crosby Group, Inc. *The Crosby Lifting Guide*. Tulsa, OK, 1998.

United States Department of Energy. *Hoisting and Rigging Manual*. Washington, DC, 1991.

United States Department of Labor. Occupational Safety and Health Administration. *OSHA 10 and 30 Hour Construction Safety and Health Outreach Training Manual*. Washington, DC, 1991.

United States Department of Labor. Occupational Safety and Health Administration. General Industry. *Code of Federal Regulations*. Title 29, Part 1910. Washington, GPO, 1998.

United States Department of Labor. Occupational Safety and Health Administration. Construction. *Code of Federal Regulations*. Title 29, Part 1926. Washington, GPO, 1998.

United States Department of Labor. Occupational Safety and Health Administration. Office of Training and Education. *OSHA Voluntary Compliance Outreach Program: Instructors Reference Manual*. Des Plaines, IL, 1993.

CHAPTER 10

HIGHWAY TRUCKS

Parked highway trucks

Materials and products of all kinds traverse U.S. highways. (See Figure 10-1.) Material handling problems do not come about during the highway portion of moving of these materials. During what is probably the shortest period of time for long-haul transportation and delivery vehicle drivers, these are the periods of loading and unloading. The danger is analogous to that of a commercial airline pilot who flies across the U.S. The pilot's most precarious time is during take-off and landing when, compared to the total flight, time is the shortest. This is the same with highway transportation. Loading and unloading take the shortest time but these are the periods when material handling is most risky.

Figure 10-1 Typical highway vehicle

Highway vehicles often require special areas and equipment for both the loading and unloading process. For example, it is expected that loading docks will be available for freight transport vehicles.

Equipment must be available to ship large loads such as fabricated materials, large steel beams, and precast concrete. Also, nonmobile equipment requires transport on flatbeds. If mobile equipment must be loaded upon flatbeds, then proper and safe loading and securing of this equipment for highway transport is essential. Loaded materials for highway transport must be fully secured and great care must be taken not to overload these highway vehicles. (See Figure 10-2.)

Figure 10-2. Secured load on highway vehicle

HIGHWAY SAFETY

Once highway vehicles are ready to hit the road, emphasis shifts from the loading and unloading process to two key components, the driver/operator and the mechanical integrity of the vehicle.

Suffice it to say that a good driver/operator decreases liability. This is why a driver's record, temperament, and physical condition need to be carefully examined to assure that potential weaknesses are detected prior to hiring, including drug and alcohol usage. Good screening is essential to a comprehensive safety and health program and to safe highway vehicle and material handling processes.

Highway vehicles must receive equal consideration since their mechanical soundness is critical to the safe handling of materials. This entails a maintenance program incorporating preventive maintenance, daily safety inspections, and immediate correction of any inadequacies or potential hazards which have been detected. Routine maintenance will pay dividends in the form of safe operations, less down time, fewer expensive on-the-road breakdowns, extended vehicle life, and the pride drivers will take in the vehicles they operate.

Once the driver is selected and the highway vehicle is properly maintained, it is time to hit the road. At this point, jurisdiction switches to the U.S. Department of Transportation (DOT) regulatory control. A driver must have a Certified Driver License (CDL) and be in compliance with other DOT regulations. But the driver is still an employee and is subject to the safety and health protection under the Occupational Safety and Health Act of 1970 (OSHAct) and the Occupational Safety and Health Administration's reporting guidelines for injuries, illnesses and deaths.

PREVENTIVE MAINTENANCE PROGRAMS

A preventive maintenance program (PMP) is geared toward heading off maintenance problems by conducting scheduled maintenance and service to prevent major breakdowns and repairs. Costwise, it will increase the longevity of vehicles and, in turn, increases the return in initial investment, accounts for less downtime, and results in increased profit affecting both employers' and operators' incomes.

A Preventive Maintenance Program (PMP) depends heavily upon a formalized approach, which is documented in writing, to insure that all vehicle and equipment inspection procedures are fully understood and accomplished and the necessary service and repairs are undertaken. It has long been noted that a PMP has benefits which extend beyond caring for equipment. Of course, equipment is expensive and if cared for properly and regularly will last a lot longer, cost less to operate, operate more efficiently, and have fewer catastrophic failures.

Properly maintained equipment is safer and there is a decreased risk of potential accidents occurring. Management's support and emphasis on having safe operating equipment will transfer to the workers in the form of better morale and respect for equipment. Well-maintained equipment sends a strong message regarding the safe operation of equipment, the value placed on employees, and about the organization's commitment to its customers. (See Figure 10-3.)

The first requirement of a PMP is to have a schedule for regular servicing and maintenance of all highway equipment; second, all defects must be reported immediately; third, supervisors must encourage all operators to conduct daily inspections; and fourth, he or she must make sure repairs are made prior to operating vehicles and equipment. If this is impossible, the equipment should be tagged and removed from service.

Figure 10-3. Highway vehicle maintenance facility

If operators are allowed to use equipment, machinery, and vehicles which are unsafe or in poor operating condition, employers send a negative message which says, "I don't value my equipment, machinery, or vehicles, and I don't value my workforce." A structured PMP will definitely foster a much more positive attitude regarding property and the workforce.

The reasons for a PMP are:

1. Improved operating efficiency of equipment, machinery, and vehicles.

2. Improved attitudes toward safety by maintaining good/safe operating equipment.

3. Fosters interaction between maintenance personnel, supervisors, and operators, and creates a degree of ownership.

4. Decreased incidents and mishaps.

A good PMP requires the following:

1. A maintenance department to carry out a regular and preventive maintenance schedule.

2. A preshift checklist for each type of equipment, machinery, or vehicle. An example of a vehicle checklist can be found in Figure 10-4.

4. Accountability and responsibility of both the supervisor and operator.

5. An effective response system when defects or hazards are discovered.

6. A commitment by you that this is important and will be done.

PRESHIFT EQUIPMENT CHECKLIST

Check any of the following defects prior to operating your equipment or vehicles and report those defects to your supervisor or maintenance department.

Type of Equipment:_____ Identification/License No.:_____

Model:_____ Year:_____ Date:_____

Walk Around:	Yes	No	N/A	Operation:	Yes	No	N/A
Broken Lights				Steering Wheel Play or Alignment			
Oil Leaks				Fuel Level			
Hydraulic Leaks				Instrument Panel			
Tires				Windshield Wipers or Washers			
Damaged Hoses				Heater or Defroster			
Bad Connections & Fittings				Mirrors			
Cracks in Windshields or Other				Seat Belts			
Glass							
Damaged Support Structures				**While Underway:**			
Damaged to Body Structures				Engine, Knocks, Misses, Overheats			
Fluid Levels:				Brakes Operate Properly			
Oil				Steering Loose, Shimmy, Hard, etc.			
Hydraulic				Transmission Noisy, Hard Shifting			
Brake				Jumps Out of Gear, etc.			
Mirrors				Speedometer			
				Speed Control			
Operation:							
Engine Starts				**Emergency Equipment:**			
Oil Pressure				First Aid Kit			
Air Pressure or Vacuum Gauge				Fire Extinguisher			
Brakes				Flags, Flares, Warning Devices			
Parking Brakes				Reflectors			
Horn				Tire Chains if Needed			
Front Lights							
Back Lights				**Cargo Related Equipment:**			
Directional Lights				Tie Downs			
Warning Lights				Cargo Nets			
Back-up Alarms				Tarps			
Noises or Malfunctioning:				Hand trucks			
Engine				Dollies			
Clutch				Ramps			
Transmission							
				Other Items:			
				Hand Tools			
				Spare Parts			

No Defects Noted: Operator Signature:_____ Date:_____ Time:_____

Describe Any Defects Noted:_____

_____

_____

Defects Corrected: Defect Correction Unnecessary: Defects Corrected By: Date:_____
(Initials)   (Initials)   (Signature)

Defects Corrected: Operator Signature   Date:_____
_____Yes _____No

Figure 10-4. Preshift Vehicle Checklist

FLEET SAFETY PROGRAM

Fleet safety may be thought of as vehicle safety or mechanical safety, but fleet safety depends upon both maintenance and operators. An operator's job includes preoperation and postoperation inspections of the vehicle and the reporting of any defects. This should be the normal operating procedure for any preventive maintenance program. Nevertheless, such expectations are dependent upon the quality and enforcement of safety policy within each organization. Operators are the true linchpins of good fleet safety programs.

Thus, in a fleet safety program, the importance of operator selection is vital to the prevention of accidents, incidents, vehicle damage, and injuries. The selection process should involve access to each operator's past employment history, driving record (including accidents), accommodations, and awards, as well as the operator's previous experience, if any, on the types of equipment for which he or she is being hired.

As a condition of employment and based upon the criteria in a written job description, all potential operators should pass a physical and mental examination and an alcohol/drug test. A well-written job description is an important and valuable tool, which in most cases can be improved when legal counsel and others to review it prior to its use.

Training is also important. All operators should undergo training related to company and government policies and procedures. This training should include recordkeeping, accident/incident reporting, driving requirements, and defensive driving. After classroom training, each operator should be required to take a supervised driving test, or hands-on supervised operational drive, to determine his/her competence. This should be done before the operator is released for work-related driving assignments. Even after the operator is released from training status, he/she may have a supervisor accompany him/her on their first work assignments.

Operators should be observed and evaluated on a periodic basis and retrained, if necessary, or supervised more closely. If you insure that your vehicles are in proper and safe operating condition, your operators become the key to your fleet safety program. Good, conscientious operators can prevent accidents from occurring; they are paramount to an effective fleet safety program.

TYPES OF HIGHWAY VEHICLES IN USE

Semitrailers for Freight

The use of semitrailers for transporting all types of packaged freight is very common place, so much so that many commercial companies have their own fleets for this purpose. Operators must make sure that their trailers are properly loaded to keep the load from shifting. Many times the operators are expected to load and unloading their own trailers. When this is the case, operators need to be provided with essential aids for moving and handling packages. Semitrailers are the most common long haul carriers upon the highways today. (See Figure 10-5.)

Tanker Semitrailers

Many liquids are transported across the U.S. highways. At times these are harmless, such as milk or orange juice, but on many occasions tankers haul very hazardous chemicals. Thus, drivers must be very aware of the dangers imposed by their cargo. If a driver is hauling a hazardous waste, a hazardous waste manifest is required and the tanker itself must be marked with the standard DOT placard and the four number identifier for the chemical. (See Figure 10-6.)

Figure 10-5. Semitrailer rigged for freight carrying

Figure 10-6. Tanker with a standard DOT placard

Drivers of tankers must be very careful during loading and unloading to prevent spills or ignition of the load. Drivers must always be alert for leaks or potential hazards to the contents of the tanker. A leak or rupture can have catastrophic effects on the environment and the individuals in close proximity who might be exposed to the tanker contents (see Figure 10-7). These types of trailers often require specially designed loading and unloading equipment.

Figure 10-7. Example of a tanker rig

Bulk Material Trailers

Bulk material such as grain, powdery solid (cement), or other fine bulk materials can be carried by bulk trailers which have the capacity to open gates on the bottom and discharge their cargo, similar to a hopper. The contents could also be solid chemicals, which require special handling and placarding. Although not as prevalent as the tanker trailer, these bulk material trailers serve an important function. (See Figure 10-8.)

Figure 10-8. Bulk handling rig

Flatbed Trailers

Flatbed trailers can be used for a myriad of activities. These include the hauling of oddly shaped materials, equipment, and specially fabricated materials. At times these loads are wider and longer than standard loads. This requires appropriate markings and escort services to add another margin of safety. Great car must be taken when using flatbed trailers to insure the load is well secured. (See Figure 10-9.)

Figure 10-9. Flatbed rig for transporting equipment

Specially Designed Trailers

Many times specially designed trailers exist to provide a special material handling function. These are unique to their particular application. (See Figure 10-10.)

Figure 10-10. Specially designed use trailer

Highway Dump Trucks

Highway dump trucks may be in the form of a long trailer or the shorter selfcontained dump truck. When dump trucks are loaded, the load should be covered to prevent discharge of the contents when in transport. The driver should remain within the cab during the loading of his/her truck or move a safe distance away in site of the individual doing the loading. In discharging the load, care should be taken to insure that the raised bed does not come in contact with electrical powerlines. The operator/driver should never place himself or herself under a raised bed unless it has been blocked from falling. Material transported by dump trucks is usually quite heavy, in the neighborhood of 3,000 pounds per cubic yard. Thus, a dump truck not under the control of the operator poses an extreme hazard. (See Figures 10-11 and 10-12.)

Figure 10-11. Dump trailer rig

Figure 10-12. The common dump truck

Delivery trucks

When considering the use of highway trucks, delivery trucks are probably the most common trucks in our cities and on our highway. The delivery truck comes in many sizes and is a single unit, not a cab and trailer truck. It is used by self-moving, package delivery, and material transport companies. It is generally not realized that such a great number of delivery trucks exist. (See Figure 10-13.)

Figure 10-3. Delivery truck

LOADING DOCKS

When discussing loading and unloading, an emphasis must be placed upon loading dock safety related to over-the-highway vehicles. Portable and powered dockboards should be strong enough to carry the load imposed on them. All portable dockboards must be secured in position, either by being anchored or equipped with devices that will prevent their slipping. Handholds, or other effective means, shall be provided on portable dockboards to permit safe handling. Positive protection must be provided to prevent railroad cars from being moved while dockboards and bridge plates are in position. (See Figure 10 -14.) A checklist for loading dock safety can be found in Appendix C.

Figure 10-14. Example of the use of dockboards for loading and unloading transport vehicles

Powered dockboards are to be designed and constructed in accordance with Commercial Standard CS202-56 (1961), "Industrial Lifts and Hinged Loading Ramps," published by the U.S. Department of Commerce, which is incorporated by reference as specified in 29 CFR 1910.6. (See Figure 10 -15.)

Figure 10-15. Powered loading dock

REFERENCES

Reese, C. D. and J.V. Eidson. *Handbook of OSHA Construction Safety & Health.* Boca Raton, FL: CRC/Lewis Publishers, 1999.

Rite Hite Corporation. *Choosing the Right Dock Leveler.* Milwaukee, WI, 1997.

Rite Hite Corporation. *Dock Design Guide.* Milwaukee, WI, 1990.

Rite Hite Corporation. *Dock Safety Guide.* Milwaukee, WI, 1989.

U.S. Department of Commerce. *Industrial Lifts and Hinged Loading Ramps.* Commercial Standard CS202-56 (1961). Washington, DC, 1961.

United States Department of Labor. Occupational Safety and Health Administration. General Industry. *Code of Federal Regulations.* Title 29, Part 1910. Washington, GPO, 1998.

CHAPTER 11

OFF-ROAD VEHICLES

Off-road vehicles

The use of off-road vehicles for material handling activities usually conjures up visions of large heavy-duty types of equipment. In many, or most, cases this is accurate. The industries that frequently utilize off-road vehicles are construction, agriculture, logging, and mining.

These vehicles often handle loads of enormous proportions, which by their sheer size and weight create a huge risk-potential of serious injury or death. This is compounded by the size of the material handling vehicle which could literally run over a worker and never know it.

The size of off-road vehicles and their loads requires enormous amounts of energy. Thus, operators and those working with or around such equipment must continuously be alert to the potential hazards presented by these operations.

OPERATORS

Operators of large material handling vehicles must be:

1. Trained to safely operate the equipment they are running.

2. Able to perform daily safety inspections and recognize defects which could cause the vehicle to operate in an unsafe manner.

3. Aware of safe operating procedures such as speed of operation, traffic patterns, and other company policies and procedures.

4. Cognizant that off-road vehicles are susceptible to many blind spots around the periphery of the vehicles.

5. Able to coordinate his/her vehicle activities with those of other off-road vehicles such as frontend loaders and dump trucks.

6. Aware of all the safety devices on his/her vehicle and can assure they are operating properly.

WORKING AROUND OFF-ROAD VEHICLES

Others working around off-road material handling vehicles must:

1. Be certain that they stay within the line of sight of the operator at all times.

2. Not position themselves between two pieces of equipment where they could be crushed.

3. Not position themselves under a load which could fall or be dropped on them.

4. Never go under a vehicle in operation or between areas where they could be pinned during the vehicle's movement.

5. Observe the swing radius of a vehicle to prevent themselves from being struck by any part of the vehicle.

6. Never work on a vehicle which has not been secured by chocks and where all potential energy has not been dissipated, locked out, or blocked to prevent unexpected release of that energy.

MOTOR VEHICLES AND MECHANIZED EQUIPMENT (29 CFR 1926.601)

The following is a summary of the OSHA construction requirements for off-highway/road vehicles. Motor vehicles which operate within off-highway jobsites, not open to public traffic, must have a service brake system, an emergency brake system, and a parking brake system. These systems may use common components and must be maintained in operable condition. Also, each vehicle must have the appropriate number of seats per occupants, and seat belts must be properly installed.

Whenever visibility conditions warrant additional light, all vehicles, or combinations of vehicles, in use are to be equipped with at least two headlights and two taillights in operable condition. All vehicles, or combination of vehicles, must have brake lights in operable condition, regardless of light conditions. Also, these vehicles shall be equipped with an adequate audible warning device at the operator's station, which is in operable condition.

No worker shall use motor vehicle equipment with an obstructed view to the rear, unless the vehicle has a reverse signal alarm, audible above the surrounding noise level, or the vehicle is only backed up when an observer signals that it is safe to do so.

All vehicles with cabs must be equipped with windshields and powered wipers. Cracked and broken glass shall be replaced. Vehicles operating in areas, or under conditions that cause fogging or frosting of the windshields, are to be equipped with operable defogging or defrosting devices.

All haulage vehicles whose payload is loaded by means of cranes, power shovels, loaders, or similar equipment must have a cab shield and/or canopy that is adequate to protect the operator from shifting or falling materials.

Tools and material are to be secured in order to prevent movement when transported in the same compartment with employees. Vehicles used to transport employees must have seats firmly secured, an adequate number of seats for the number of employees carried, and seat belts for everyone.

Trucks with dump bodies shall be equipped with positive means of support, permanently attached, and capable of being locked in position to prevent accidental lowering of the body while maintenance and inspection work is being done. Operating levers, controlling, hoisting and dumping devices on haulage bodies must be equipped with a latch or other device, to prevent accidental starting or tripping of the mechanism. Trip handles for tailgates of dump trucks shall be so arranged that, in dumping, the operator will be in the clear. All rubber-tired motor vehicle equipment is to be equipped with fenders. Mud flaps are to be used in lieu of fenders whenever motor vehicle equipment is not designed for fenders.

All vehicles in use should be checked at the beginning of each shift to assure that the following parts, equipment, and accessories are in safe operating condition and free of apparent damage that could cause failure while in use: service brakes, including trailer brake connections; parking system (hand brake); emergency stopping system (brakes); tires; horn; steering mechanism; coupling devices; seat belts; operating controls; and safety devices. All defects are to be corrected before the vehicle is placed in service. These requirements also apply to equipment such as lights, reflectors, windshield wipers, defrosters, fire extinguishers, etc., where such equipment is necessary.

All equipment left unattended at night adjacent to highways or construction areas must have lights, reflectors, and/or barricades to identify location of the equipment. Supervisory personnel must ensure that all machinery and equipment are inspected prior to each use and verify that both are in safe operating condition. Rated load capacities and recommended rules of operation must be conspicuously posted on all equipment at the operator's station. An accessible fire extinguisher of 5-BC rating or higher must be available at all operator stations. When vehicles or mobile equipment are stopped or parked, the parking brake must be set. Equipment on inclines must have wheels chocked, as well as the parking brake set.

MATERIAL HANDLING EQUIPMENT (29 CFR 1926.602).

This is a continuation of the construction requirements for off-road vehicles, which perform material handling tasks and include the following types of earthmoving equipment: scrapers, loaders, crawlers or wheel tractors, bulldozers, off-highway trucks, graders, agricultural and industrial tractors, and similar equipment. (See Figure 11-1.)

Figure 11-1. Example of material handling equipment

Seat belts are to be provided on all equipment covered by this section of OSHA's regulations, and they must meet the requirements set by the Society of Automotive Engineers, J386-1969, "Seat Belts for Construction Equipment." Seat belts for agricultural and light industrial tractors must meet the seat belt requirements of Society of Automotive Engineers J333a-1970, "Operator Protection for Agricultural and Light Industrial Tractors." Seat belts are not needed for equipment designed for only standup operation. Seat belts are not needed for equipment that does not have Rollover Protective System (ROPS) or adequate canopy protection.

No construction equipment or vehicles are to be moved upon any access roadway or grade unless the access roadway or grade is constructed and maintained to accommodate safe movement of the equipment and vehicles involved. Every emergency access ramp and berm used by an employer is to be constructed to restrain and control runaway vehicles. All earth-moving equipment shall have a service braking system capable of stopping and holding the equipment fully loaded, as specified in Society of Automotive Engineers. Brake systems for self-propelled rubber-tired off-highway equipment must meet the applicable minimum performance criteria set forth in the following Society of Automotive Engineers Recommended Practices:

SAE J319b-1971—Self-Propelled Scrapers;
SAE J236-1971—Self-Propelled Graders;
SAE J166-1971—Trucks and Wagons;
SAE J237-1971—Front End Loaders and Dozers.

Pneumatic-tired earth-moving haulage equipment (trucks, scrapers, tractors, and trailing units) whose maximum speed exceeds 15 miles per hour is to be equipped with fenders on all wheels, unless the employer can demonstrate that uncovered wheels present no hazard to personnel from flying materials. Rollover protective structures should be found on all material handling equipment. (See Figure 11-2.)

Figure 11-2. A scraper with rollover protection

All bi-directional machines, such as rollers, compactors, front-end loaders, bulldozers, and similar equipment, shall be equipped with a horn, distinguishable from the surrounding noise level, that must be operated, as needed, when the machine is moving in either direction. The horn shall be maintained in an operative condition. No earth-moving or compacting equipment which has an obstructed view to the rear is permitted to be used in the reverse gear unless the equipment has in operation a reverse signal alarm which is distinguishable from the surrounding noise level, or there is an authorized employee who signals that it is safe to do so. (See Figure 11-3.)

Figure 11-3. Front end loaders are bi-directional and required to have a back-up alarm

Scissor points on all front-end loaders, which constitute a hazard to the operator during normal operation, are to be guarded.

Tractors must have seat belts. They are required for operators when they are seated in the normal seating arrangement for tractor operation. When back-hoes, breakers, or other similar attachments are used on tractors for excavating or other work, which requires other than normal seating, seat belts must be worn.

Industrial trucks, such as lift trucks, forklifts, stackers, etc., must have the rated capacity clearly posted on the vehicle and clearly visible to the operator. When auxiliary removable counterweights are provided by the manufacturer, corresponding alternate rated capacities also must be clearly shown on the vehicle. These ratings shall not be exceeded. No modifications or additions which affect the capacity or safe operation of the equipment shall be made without the manufacturer's written approval. If such modifications or changes are made, the capacity, operation, and maintenance instruction plates, tags, or decals must be changed accordingly. In no case shall the original safety factor of the equipment be reduced. If a load is lifted by two or more trucks working in unison, the proportion of the total load carried by any one truck must not exceed its capacity. Steering or spinner knobs are not to be attached to the steering wheel unless the steering mechanism is of a type that prevents road reactions from causing the steering handwheel to spin. The steering knob shall be mounted within the periphery of the wheel.

All high lift rider industrial trucks are to be equipped with overhead guards which meet the configuration and structural requirements as defined in paragraph 421 of American National Standards Institute B56.1-1969, "Safety Standards for Powered Industrial Trucks." All industrial trucks in use shall follow the applicable requirements of design, construction, stability, inspection, testing, maintenance, and operation, as defined in American National Standards Institute B56.1-1969.

Unauthorized personnel are not permitted to ride on powered industrial trucks. A safe place to ride is to be provided when riding on trucks is authorized. Whenever a truck is equipped with vertical only, or vertical and horizontal controls, elevatable with the lifting, carriage, or forks for lifting personnel, the following additional precautions are to be taken for the protection of the personnel being elevated. They are to use a safety platform, which is firmly secured to the lifting carriage and/or forks, and a means must be provided whereby personnel on the platform can shut off power to the truck. Personnel are also to be protected from falling objects, if falling objects could occur due to the operating conditions.

MATERIAL HANDLING EQUIPMENT SAFETY, TRAINING, AND INSPECTION

The best practices for the training of drivers of off-road equipment do not constitute a complete training program. They should include adherence to safety practices and vehicle inspections. This process will provide vehicle drivers with safer operating equipment. The following list of best practices gives the driver a starting kit of the best tools, which some of the safest companies in the U.S. have put together to help keep the driver safe. In training employees to use equipment be aware that one size does not fit all. There are possible differences that can prohibit use of some of the best practices listed here at your specific operation.

Proper training of a truck driver is of the utmost importance. Training should always be done before the driver starts production work. If possible, all new drivers should be tested on their knowledge after their initial hands-on training, then rechecked periodically.

At intervals, drivers should be observed to ensure that no bad habits have developed, and to reinforce the training. Subsequent training should be conducted as needed. Training should always be renewed as an opportunity to develop a workforce into the safest, most cost-effective, efficient, and productive team possible. It is up to employers and managers to provide the opportunity to make this happen.

General Safety

<u>Personal Safety Equipment</u>

- Hard hat, steel toe boots, safety glasses, hearing protection, and dust protection.

<u>Seat Belts</u>

- Seat and seat belt are in good working order.
- Seat belts are required at all times when vehicle is in use.

<u>Pre-Operation Inspection</u>

Where applicable, all of the items below should be completed on every pre-operation inspection. Make sure the vehicle is in a safe location prior to making the pre-operational check.

- Fluid levels—engine oil, hydraulic oil, steering oil, brake oil, coolant fluid, and fuel.
- Steering components.
- Tires, lug nuts, wheels, and flanges.
- Frame and bed for cracks and damages.
- Visibility systems—mirrors, cameras, windows, windshield wipers, etc.
- Power train—engine, torque converter, transmission, differential, and final drive.
- Electric drive alternators and wheel motors.
- Hoist cylinders—mounts, pins, and pin keepers.
- Nose cone/wish bone assembly—cracks, proper lubrication and looseness.
- Brake test—make sure all brakes hold to manufacturers' specifications.
- Warning devices are functional—gauges, lights, buzzers, and backup alarm.
- Fire suppression system/extinguishers—pins & keepers in place, tags current, hoses, etc.
- Wheel chocks available for use.
- Ladders, handrails, and steps are accessible.
- Headlights, clearance, turn-signals, taillights, and brake lights.
- Heaters and defrosters function properly.
- Cab doors open and close properly.
- Operators manual easily available.
- Air pressure system works properly.
- Belts and guards.
- Radio.

Vehicle Operation

The operator should follow the following operational steps every day for every vehicle.

Vehicles Operation

1. Sit in an upright position with the seat belt fastened at all times.
2. Always maintain a safe and reasonable rate of speed and following distance, considering the road conditions, weather, traffic, and load.
3. Test all braking systems to insure proper function before operating vehicle.
4. Check operators manual for correct procedure and other limitations for your vehicle to accomplish the task at hand.
5. Pass other vehicles only when adequate clearance and visibility is present and communicate your intentions.
6. Do not pass at intersections.
7. Avoid running over rocks or into potholes or ruts.
8. Test all steering functions prior to operation.
9. Secure loose items in the cab.

Know The Controls (Location and Operation)—All Brakes, Signals, Accessories, Instrumentation and Warning Devices

1. How they work, normal and abnormal readings, and what should be done if an alarm sounds.
    (All international symbols should be explained to the operator.)

Proper Start-Up and Shut-Down Procedures

It is most important that operators of off-road vehicles follow the proper start-up and shut-down procedures when these types of vehicles are to be used.

1. Before starting the engine, ensure that all is clear. (Do not proceed if visibility is impaired.
2. Warn others before moving (honk horn).
3. Warn others before exiting ready-line (honk horn).
4. Follow prescribed procedures for cold or warm engine starting.
5. Allow time to warm up before operating.
6. Choose a safe location to park your vehicle.
7. Allow time to cool down before shut-down.
8. Set park brake and turn off lights.
9. Set wheel chocks if necessary.

Working Procedures

Operator's Responsibilities

1. Safe, productive operation of the equipment with a minimum amount of preventable downtime due to mechanical failure.

2.   The elimination of property damage and accidents by using care and consideration around other equipment and operators.

3.   Report any unsafe conditions immediately.

<u>Speed Control</u>

1.   Throttle, retarder, and brakes.

Haulage Trucks

Off-road haulage trucks face many inherent problems that are not experienced by operators of highway dump trucks even though many similarities exist. (See Figure 11-4.)

Figure 11-4. Example of an off-road haulage truck on the left

<u>Spotting At Loading Equipment (For Haulage Trucks)</u>

1.   Check clearances.

2.   Visually check loading area on approach to be sure that no equipment or workers are behind the truck before reversing.

3.   Pay close attention to areas with steep drop-offs.

4.   Watch closely for other equipment, persons, small vehicles, etc.

<u>Spotting At Dump Locations</u>

1.   Check approach berm height/thickness. (Reminder: when backing up to a dump, use the berm as a guide only.)

2.   Look for cracked ground/settling/bulges.

3.   Report any unsafe conditions immediately to the supervisor and other drivers.

4.   If spotters are provided, have direct communication with them.

Operating On Grades

1.   Apply retarder and reduce speed prior to descending grades.

2.   Gear selection. (Note: On mechanical drive trucks it is very important for the operator to slow the truck down prior to descending the grade and place the truck in the gear corresponding with the grade profile being operated on and the manufacturer's specifications. The gear to be used will vary between truck types and the percent of grade used. Always follow the grade profile charts provided by the manufacturer in the operator's manual. Improper gear selection may cause excessive brake temperature, low brake air pressures, and reduced steering oil pressures.)

3.   Use retarder to maintain proper speed.

4.   Electric drive units—check for proper voltage prior to descending grade. (Note: To ensure dynamic braking on electric drive trucks, the operator must check before starting down the grade to see that proper voltage is being supplied to the wheel motors. This voltage will vary between truck types, so the manufacturer's specifications must be checked and adhered to. Improper voltage will cause loss of dynamic braking.)

Right-of-way Procedures

1.   Drive defensively and follow established traffic patterns and controls.
2.   Loaded truck generally has right-of-way.
3.   When in doubt, YIELD!

Reduction Of Component Damage

1.   Engine—proper RPM, oil pressure, and coolant temperature

2.   Tires—avoid rocks in the road. Know where the blind side of the truck is

3.   Proper speeds and loads.

4.   Reduce speed when turning or traveling over rough terrain.

Machine Systems

The driver should know the difference between proper operation and possible malfunction. The operator should understand how the systems work and know the different components that make up the systems. Operators should have knowledge of:

- Brake systems.
- Steering systems.
- Drive train.
- Warning systems.
- Accessories.

- Emergency shut-down procedures.
- Fire-suppression system.

Housekeeping

1. Keep vehicles free of combustible materials (oil, grease, etc.).

2. All loose items in cab should be secured to prevent motion at all times.

3. Keep ladders, walkways, and cabs clear of extraneous materials and tripping hazards.

It is important that every vehicle operator read the operators manual and use good common sense when operating any piece of equipment. Always report any changes, such as unusual sounds and operating responses in your equipment (anything your senses can pick up). Figure 11-5 has a pre-operation checklist for off-road vehicle operators.

DATE:_____

VEHICLE #_____

OPERATOR:_____

SHIFT: _____

PRE-OPERATION INSPECTION—CHECK BEFORE OPERATING

CATEGORY	OK	NOT OK	COMMENTS
Seat Belts			
Back-up Alarm			
Brakes (service/retarder, secondary, park)			
Low Air Pressure			
Steering			
Speedometer			
Fire Suppression System			
Emergency Monitoring System Stage Alarm			
Tires flats & lug nuts loose			
Pins (hoist cylinder and body w/retainers)			
Auto Lube System			
Fluid Leaks (fuel, coolant, engine oil, transmission oil)			
hydraulic oil, steering, converter, and brakes)			
Lights (head, tail, brake, retarder, clearance. Hazard, panel)			
Fire Extinguisher (portable)			
Glass / Mirrors (circle)			
Horn			
Wheel Chocks / Lunch Pail Rope (circle)			

Figure 11-5. Pre-operation off-road vehicle checklist

CATEGORY	OK	NOT OK	COMMENTS
Windshield Wipers			
Heater / Air Conditioner (circle)			
Rock Ejectors			
Grab Irons/Steps / Ladders (circle)			
Frame Cracks / Bed Cracks / Nose Cone Assembly (circle)			
Clean Working Place			
Drain Air Tanks (main, secondary, governor)			
Operator's Seat / Passengers Seat (circle)			
Suspensions			
Doors			
Safety Chains & Cables			
Exhaust System			
Air Cleaners Plugged			
Hoist Cylinders (hard to dump)			
Canopy & Rock Guards			
Radio			
Emergency Monitoring System (test)			
Payload Monitoring System (list any fault codes)			
Automatic Electronic Traction Aid System			
Ducktail on Bed Intact			
Does Automatic Retarder Work Correctly			
Fuel Level			
ELECTRICAL:			
Slow Going Uphill Loaded			
Speedometer or Tachometer			
Gauges and All Other Warning Devices (overspeed monitor. etc.)			
Computer or Computer light			
Starter			
Switches			

Immediately report any condition which will result in unsafe operation of this vehicle to your supervisor.

Circle when and if repaired.

Figure 11-5. Pre-operation off-road vehicle checklist (Continued)

Rollover protection should be an integral component of all off-road vehicles which are not designed for stand up operations. (See Figure 11-6.)

Figure 11-6. Example of rollover protection on a dozer

REFERENCES

Reese, C. D. and J.V. Eidson. *Handbook of OSHA Construction Safety & Health.* Boca Raton, FL: CRC/Lewis Publishers, 1999.

United States Department of Labor. Mine Safety and Health Administration. *MSHA Special Emphasis Program on Off-Road Haulage.* Arlington, VA, 1997.

United States Department of Labor. Occupational Safety and Health Administration. *OSHA 10 and 30 Hour Construction Safety and Health Outreach Training Manual.* Washington, DC, 1991.

United States Department of Labor. Occupational Safety and Health Administration. Construction. *Code of Federal Regulations.* Title 29, Part 1926. Washington, GPO, 1998.

CHAPTER 12

INDUSTRIAL TRUCKS (FORKLIFTS)

Powered industrial truck

Powered industrial trucks (forklifts) are among the most useful and important material handling vehicle within the workplace or jobsite. In recent years we have become very aware that the misuse of this type of lifting vehicle has resulted in many injuries and deaths. Thus, special precautions and driver training are of the utmost importance in the safe use of powered industrial trucks.

INCIDENCE OF LIFT-TRUCK INJURIES

Each year, it is estimated that more than 95,000 powered lift-truck-related injuries and 100 deaths (see Figure 12-1) occur in U.S. industry (OSHA 1999). Injuries involve em-

179

ployees being struck by lift trucks or falling while standing/working from elevated pallets and tines. Many employees are injured when lift trucks are inadvertently driven off loading docks or when the lift falls between a dock and an unchocked trailer. For each employee injured, there are probably numerous incidents that are unnoticed or unreported to supervisors. All mishaps, no matter how small, are costly. Most incidents also involve property damage. Damage to overhead sprinklers, racking, pipes, walls, machinery, and various other equipment occurs all too often. In fact, million of dollars are lost in damaged equipment, destroyed products, or missed shipments. Unfortunately, a majority of employee injuries and property damage can be attributed to lack of procedures, insufficient or inadequate training, and lack of safety-rule enforcement.

How accident occurred	Number	Percent
Forklift overturned	41	24
Forklift struck something, or ran off dock	13	8
Worker pinned between objects	19	11
Worker struck by material	29	17
Worker struck by forklift	24	14
Worker fell from forklift	24	14
Worker died during forklift repair	10	6
Other accident	10	6
Total	170	100

Source: Bureau of Labor and Statistics, Fatal Workplace Injuries in 1992, A Collection of Data and Analysis, Report 870, April 1994.

Figure 12-1. Classification of forklift fatalities, 1991-1992

Unsafe Acts and Conditions

Some examples of the unsafe acts and conditions that occur during the use of powered industrial trucks are as follows:

- Unsafe Acts
  - Inadequately trained maintenance personnel, inspectors and operators.
  - Wrong truck selected for the job (too big, too small, wrong for hazardous location).
  - Hurrying, taking shortcuts, not paying attention, fatigue, boredom, or not following the rules.
  - Overloading trucks.
  - Improper selection and installation of dockboards and bridge plates.
- Unsafe Conditions
  - Forks or other load-handling attachments cracked or bent.
  - Gouges or large chunks missing from solid tires.
  - Blind corners.

- Leaky connectors and hydraulic cylinders.
- Too much free play in the steering.
- Unsafe refueling or recharging practices.

HAZARDS AND EFFECTS

Many hazards associated with the operation of powered industrial trucks are the result of common operator mistakes. For instance, collisions between trucks and stationary objects often occur while trucks are backing up — usually while turning and maneuvering. Unless care is exercised, operators can cause damage to overhead fixtures (e.g., sprinklers, piping, electrical conduits) while traveling and maneuvering under them.

Accidents often occur when an operator leaves a truck so that it obstructs a passageway and an unauthorized (untrained) worker tries to move it. Other common hazards include carrying unstable loads, tipping over trucks, dropping loads on operators or others, running into or over others, and pinning others between the truck and fixed objects.

Unauthorized passengers are often seriously injured from falling off trucks. Unless space is provided, do not allow passengers to ride on the trucks.

Dangerous misuse of trucks includes bumping skids, moving piles of material out of the way, moving heavy objects by means of makeshift connections, and pushing other trucks. All these activities can cause accidents or injuries; they also indicate poor operator training.

Factors that can influence stability (resistance to overturning) must be considered. These include:

- Weight, weight distribution, wheel base, tire tread, truck speed, and mast defection under load.
- Improper operation, faulty maintenance, and poor housekeeping.
- Ground and floor conditions, grade, speed and judgment of the operator.

PREVENTION OVERVIEW

Whether the operator is new to the job or experienced, he or she should visually check forklift trucks every day. Good prevention consists mainly of proper maintenance, trained operators and adherence to established safety procedures. Special attention should be given to the following areas:

- Proper truck selection (size, load-carrying capacity, hazardous locations).
- Condition and inflation of pressure lines.
- Battery, lights, and warning devices.
- Controls, including lift and tilt system and limit switches.
- Brakes and steering mechanisms.
- Fuel system.

TYPES OF POWERED INDUSTRIAL TRUCKS

These general requirements for powered industrial trucks contain safety requirements related to fire protection, design, maintenance, and the use of fork trucks, tractors, platform lift trucks, motorized hand trucks, and other specialized industrial trucks powered by electric motors or internal combustion engines. These requirements do not apply to compressed air or non-flammable compressed gas-operated industrial trucks, nor to farm vehicles, or to vehicles intended primarily for earth moving or over-the-road hauling.

Approved powered industrial trucks shall bear a label or some other identifying mark indicating approval by the testing laboratory. Modifications and additions which affect capacity and safe operation of these trucks shall not be performed by the user without manufacturers' prior written approval.

As used in this chapter, the term "approved truck" or "approved industrial trucks" means a truck that is listed or approved for fire safety purposes for the intended use by a nationally recognized testing laboratory, using nationally recognized testing standards.

There are eleven different designations of industrial trucks or tractors, which are D, DS, DY, E, ES, EE, EX, G, GS, LP, and LPS. The meaning of each of these is provided as follows:

1. "D" designated units are diesel engine powered units having minimum acceptable safeguards against inherent fire hazards.

2. "DS" designated units are diesel powered units that are provided with additional safeguards to the exhaust, fuel, and electrical systems.

3. "DY" designated units are diesel powered units that have all the safeguards of the "DS" units and, in addition, do not have any electrical equipment including the ignition and are equipped with temperature limitation features.

4. "E" designated units are electrically powered units that have minimum acceptable safeguards against inherent fire hazards.

5. "ES" designated units are electrically powered units that, in addition to all the requirements for the "E" units, are provided with additional safeguards to the electrical system to prevent emission of hazardous sparks and to limit surface temperatures.

6. "EE" designated units are electrically powered units that have, in addition to all of the requirements for the "E" and "ES" units, the electric motors and all other electrical equipment completely enclosed.

7. "EX" designated units are electrically powered units that differ from the "E," "ES," or "EE" units in that the electrical fittings and equipment are so designed, constructed and assembled so that the units may be used in certain atmospheres containing flammable vapors or dusts.

8. "G" designated units are gasoline powered units having minimum acceptable safeguards against inherent fire hazards.

9. "GS" designated units are gasoline powered units that are provided with additional safeguards to the exhaust, fuel, and electrical systems.

10. "LP" designated unit is similar to the "G" unit except that liquefied petroleum gas is used for fuel instead of gasoline.

11. "LPS" designated units are liquefied petroleum gas powered units that are provided with additional safeguards to the exhaust, fuel, and electrical systems.

Atmospheres or locations throughout the plant must be classified hazardous or non-hazardous prior to the consideration of industrial trucks being used therein. Refer to Table 12-1 and 12-2, which is a summary table on use of industrial trucks in various locations.

Table 12-1

Hazardous Locations Classifications and Groups

Courtesy of U.S. Department of Energy

Classes	Unclassified	Class I locations	Class II locations	Class III locations
Description of Classes	Locations not possessing environments described in other columns.	Locations in which flammable gases or vapors are, or may be, present in the air in quantities sufficient to produce explosive or ignitable mixtures.	Locations that are hazardous because of the presence of combustible dust.	Locations in which easily ignitable fibers are present but not likely to be in suspension in quantities sufficient to produce ignitable mixtures.

Groups within classes	None	A	B	C	D	E	F	G	None
Examples of locations or environments in classes and groups	Piers and wharves inside and outside general storage, general industrial or commercial properties	Acetylene	Hydrogen	Ethyl ether	Gasoline Naphtha Alcohols Acetone Lacquer Solvent Benzene	Metal dust	Carbon black, coal dust, coke dust	Grain dust, flour dust, starch dust, organic dust	Baled waste, cocoa fiber, cotton, excelsior, hemp, istle, jute, kapok, oakum, sisal, Spanish moss, fibers, tow

Table 12-2

Authorized Uses of Trucks by Types in Groups of Classes and Divisions

Courtesy of U.S. Department of Energy

Groups within classes	None	A	B	C	D	A	B	C	D	E	F	G	E	F	G	None	None
Type of truck authorized																	
Diesel																	
Type D	D**																
Type DS									DS			DS				DS	
Type DY									DY			DY	DY			DY	
Electric																	
Type E	E**															E	
Type ES									ES			ES				ES	
Type EE									EE			EE	EE			EE	
Type EX				EX					EX	EX	EX	EX	EX			EX	
Gasoline																	
Type G	G**																
Type GS									GS			GS					GS
LP-Gas																	
Type LP	LP**																
Type LPS									LPS			LPS					LPS
*Paragraph Ref. in No. 505	210, 211		201(a)		203(a)	209(a)			204(a) 204(b)	202(a)	205(a)	209(a)	206(a) 206(b)	207(a)	208(a)		

*See NFPA No. 505-1969, Powered Industrial Trucks.

**Trucks conforming to these types may also be used.

PROTECTIVE DEVICES

The use of protective devices is an important factor in safe forklift operation. Safety specialists can assist supervisors in determining what protective devices are necessary. Although forklifts need not be equipped alike, there are some similarities, such as lights. Also, manufacturers are required by federal standards to equip forklifts with certain mandatory features, such as back-up alarms. When a truck is about to move in reverse, it is required to sound a warning. Some other protective devices include:

- Overhead protection to guard the operator from falling objects.
- Wheel plates to protect the operator from objects picked up and thrown by tires.
- On-board fire extinguishers.
- Horns to warn others when the truck is moving forward.

Other protection devices that might be seen in the work area, which are specifically designed for the operator, include:

- Signs—such as stop, caution, danger, and speed limits—to inform operators of conditions.
- Gloves and safety shoes.
- Eyewash stations.
- Concave mirrors.
- Eye protection devices.
- Hardhats to protect operators when there is an overhead hazard.

WORK PRACTICES

Selection and Inspection of Trucks

Industrial trucks shall be examined before being placed in service, and shall not be placed in service if an examination shows any condition adversely affecting the safety of the vehicle. Examinations shall be made at least daily. Where trucks are used on a round-the-clock basis, they shall be examined after each shift. Figure 12-2 shows the major components of a standard forklift.

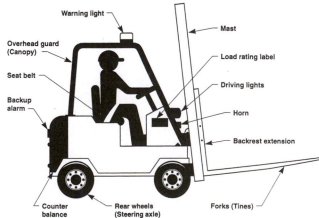

Figure 12- 2. Components of a forklift. Courtesy of the U.S. Department of Energy

The proper truck (size, load capacity, and use) must be selected and inspected to ensure that all controls and other safety features are working properly. All powered industrial truck operators must check the vehicle which they are operating at the start of each shift. If the vehicle is found to be unsafe, it must be reported to the manager immediately. No powered industrial truck should be operated in an unsafe condition. It is a good idea to use and maintain a daily preshift checklist to monitor the condition of powered industrial truck (forklifts). The operating condition of a forklift may change throughout each day and between shifts. An inspection identifies potential hazards both prior to operation and at the end of use of the powered industrial truck. Attention shall be given to the proper functioning of tires, horns, lights, battery, controller, brakes, steering mechanism, and the lift system of fork lifts (fork chains, cable, and limit switches). Special attention should be given to the following:

1. Prior to initial use, all new, altered, modified, or extensively repaired forklifts shall be inspected by a qualified inspector to insure compliance with the provisions of this book.

2. Brakes, steering mechanisms, control mechanisms, warning devices, lights, governors, lift overload devices, guards, and other safety devices shall be inspected regularly and maintained in a safe-operating condition.

3. All parts of the lift and tilt mechanisms and frame members must be carefully and regularly inspected and maintained in a safe-operating condition.

4. Special trucks or devices designed and approved for operation in hazardous areas shall receive special attention to ensure that the original, approved safe-operating features are preserved by maintenance.

5. Fuel systems shall be checked for leaks and condition of parts. Extra-special consideration must be given in the case of a leak in the fuel system. Action shall be taken to prevent the use of the truck until the leak has been corrected.

6. All hydraulic systems must be regularly inspected and maintained in conformance with good practice. Tilt cylinders, valves, and other similar parts shall be checked to assure that "drift" has not developed to the extent that it would create a hazard.

7. Capacity, operation, and maintenance-instruction plates, tags, and decals must be maintained in a legible condition.

8. Batteries, motors, controllers, limit switches, protective devices, electrical conductors, and connections shall be inspected and maintained in conformance with good practice. Special attention must be paid to the condition of electrical insulation.

9. Inspect the mast for broken or cracked weld-points and any other obvious damage.

10. Make sure roller tracks are greased and that chains are free to travel.

11. Be sure the forks are equally spaced and free from cracks along the blade and at the heels.

12. Check tires for excessive wear, splitting, or missing tire material, as well as inflation levels.

13. If a powered industrial truck (forklift) is powered by propane, inspect the tank for cracks, broken weld-points, and other damage. Make sure all valves, nozzles and hoses are secure and do not leak.

Once you have completed the inspection and maintenance, the operator should then get in the seat to check:

- Brakes.

- Oil pressure gauge, water temperate gauge.

- Steering. (The wheel should turn correctly in both directions.)

- Operation of the headlights, taillights and warning lights.

- Clutch.

- Back up alarm.

Maintenance and Repair of Trucks

It is required that trained and authorized personnel maintain and inspect the powered (forklift) industrial trucks. All work should be done in accordance with the manufacturer's specifications. Because of everyday use of these vehicles, it is particularly important for personnel to overhead follow the maintenance, lubrication, and inspection schedules. Special attention should be given to forklift control and lifting features, such as brakes, steering, lift apparatus, overload devices, and tilt mechanism.

Any power-operated industrial truck not in safe operating condition shall be removed from service. All repairs must be made by authorized personnel. No repairs shall be made in Class I, II, or III locations. Repairs to the fuel and ignition systems, which involve fire hazards, must be conducted only in locations designated for such repairs.

Changing and Charging Storage Batteries

Workplaces using electrically powered industrial trucks will have battery-charging areas somewhere in the plant. In many cases, depending on the number of electrically powered industrial trucks, there will be more than one changing and charging area. This section only applies to storage battery changing and charging areas associated with powered industrial trucks. It does not apply to areas where other batteries, such as those used in motor vehicles (cars or trucks), are charged, although some of the same hazardous conditions may exist. Some of the requirements specified in the regulation include:

1. Make sure batteries are checked for cracks or holes, security sealed cells, frayed cables, broken insulation, tight connections, and clogged vent caps.

2. Battery charging installations shall be located in areas designated for that purpose.

3. Facilities must be provided for flushing and neutralizing spilled electrolyte, for fire protection, for protecting charging apparatus from damage by trucks, and for adequate ventilation for dispersal of air contaminants from gassing batteries.

4. A conveyor, overhead hoist, or equivalent material handling equipment shall be provided for handling batteries.

5. Smoking shall be prohibited in the charging area.

6. Precautions shall be taken to prevent open flames, sparks, or electric arcs in battery charging areas.

The Rated Capacity

Rated capacity is the maximum weight that a powered industrial truck can transport and stack at a specified load center and for a specified load elevation. When originally purchased, this is usually the maximum weight, expressed in kilograms (pounds) of a 1,200 milli-meters (48 inches) homogenous cube (600 mm load center) that a truck can transport and stack to a height established by the manufacturer. Industrial trucks shall not be used or tested above their special rated capacity. (See ANSI/ASME B56.1)

Load Testing

Forklifts shall be load tested and inspected by a qualified inspector when assigned to service, and thereafter at 12-month intervals. Load test records shall be kept on file and readily available to appointed personnel. The load tests required must not exceed the rated capacity of the equipment. Test weights shall be accurate to within five percent plus zero percent of stipu-lated values. Load slippage for this equipment must not be greater than a maximum of three inches vertically and one inch horizontally at the cylinder during a static test period of at least ten minutes in duration. If a test has not been completed by the end of the required period, the equipment shall be downrated as follows:

1. Thirty calendar days after the end of the period, the equipment shall be downrated to 75 percent of the rated capacity.

2. Sixty calendar days after the end of the period, the equipment shall be downrated to 50 percent of the rated capacity.

3. Ninety calendar days after the end of the period, the equipment shall be taken out of service until the required inspection has been completed.

Industrial Truck Nameplate

Every powered industrial truck shall have appended to it a durable, corrosion-resis-tant nameplate with the model or serial number and appropriate weight of the truck legibly inscribed. The serial number shall also be stamped on the frame of the truck. The truck must be accepted by a recognized national testing laboratory and the nameplate shall be marked. The truck shall meet all other nameplate requirements of ANSI/ASME B56.1.

Every removable attachment (excluding fork extensions) must have installed a du-rable corrosion-resistant nameplate with the following information legibly and permanently inscribed:

- Serial number.

- Weight of attachment.

- Rated capacity of attachment.

- The following instructions (or equivalent): "Capacity of truck and attachments combination may be less than capacity shown on attachment—consult truck name-plate."

SAFETY TIPS FOR OPERATING POWERED INDUSTRIAL TRUCKS

Safe Operations

Operators must follow all safety rules related to speed, parking, fueling, loading, and moving loads. While the forklift is in operation, keep the forks low with the mast tilted slightly back. Too tall or "top-heavy" loads can change the forklift's center of gravity and cause it to tip over. Follow safe speed limits. Loaded forklifts should travel at low speeds. Without loads, forklifts are not weighted and are especially unstable. Avoid sharp turns. Forklifts can turn over if turns are made too fast. When parking on a hill, always chock the forklift's wheels, lower the tines, and set the parking brake. Also, to avoid tipping, always carry loads up a grade and back down ramps. Never turn on grades. Keep safe visibility. If a load blocks forward vision, drive backwards. Always use the horn at intersections. Be cautious around uneven surfaces; chuckholes, and other uneven ground can cause forklifts to tip. The following are some general safety rules for operating a powered industrial truck:

1.   Only drivers authorized by the company and trained in the safe operation of forklift trucks or pickers shall be permitted to operate such vehicles. Drivers may not operate trucks other than those for which they are authorized.

2.   Drivers must check the vehicle at least once per day and if it is found to be unsafe, the matter shall be reported immediately to a manager or mechanic, and the vehicle may not be put into service again until it has been made safe.

3.   No person shall be allowed to stand or pass under the elevated portion of any truck, whether loaded or empty.

4.   Unauthorized personnel may not be permitted to ride on powered industrial trucks. A safe place to ride shall be provided where riding of trucks is authorized.

5.   When a powered industrial truck is left unattended, load-engaging means shall be fully lowered, controls must be neutralized, power shall be shut off, and brakes set. Wheels shall be blocked if the truck is parked on an incline. A powered industrial truck is "unattended" when the operator is 25 feet or more away from the vehicle which remains in his view, or whenever the operator leaves the vehicle and it is not in his view.

6.   When the operator dismounts and is within 25 feet of the truck still in his or her view, the load-engaging means shall be fully lowered, control neutralized, and the brakes set to prevent movement. (See Figure 12-3.)

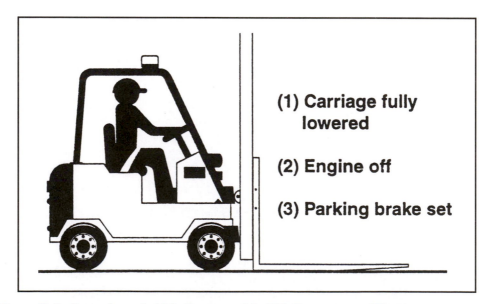

Figure 12-3. Properly set forklift. Courtesy of the U.S. Department of Energy

7. The vehicle shall not exceed the authorized or safe speed, must always maintain a safe distance from other vehicles, and must observe all established traffic regulations. For trucks traveling in the same direction, a safe distance may be considered to be approximately three truck lengths or, preferably, a time lapse of three seconds before passing the same point. Exercise extreme care when cornering. Sound horn at blind corners.

8. Employees shall not place any part of their bodies outside the running lines of the forklift truck or between mast uprights or other parts of the truck where shear or crushing hazards exits.

9. The width of one tire on the forklift should be the minimum distance maintained by the truck from the edge while it is on any elevated dock, platform or freight car.

10. Stunt driving and horseplay are prohibited.

11. Trucks shall not be loaded in excess of their rated capacity.

12. Extreme care must be used when lifting loads, and loaded vehicles shall not be moved until the load is safe and secure.

13. Extreme care should be taken when tilting loads. Elevated loads shall not be tilted forward except when the load is being deposited onto a storage rack or equivalent. When stacking or tiering, backward tilts shall be limited to that which is necessary to stabilize the load.

14. Operators must look in the direction of travel and shall not move a vehicle until certain that all persons are in the clear.

15. Vehicles shall not be operated on floors, sidewalk doors, or platforms that will not safely support the vehicle, empty or loaded. Any damage to forklift trucks and/or structures must be reported immediately to the manager. Additionally, doors adjacent to the path of vehicles should be marked and secured where possible.

16. The forks shall always be carried as low as possible, consistent with safe operation.

17. Special precautions must be taken in the securing and handling of loads by trucks equipped with attachments, and during the operation of these trucks after the loads have been removed.

18. Vehicles shall not be driven in and out of highway trucks and trailers at unloading docks until such trucks are securely blocked and brakes set.

19. No truck should operate with a leak in the fuel system.

20. The load-engaging device must be place in such a manner that the load will be securely held or supported.

21. No smoking is permitted while operating or refueling forklifts.

22. A fire extinguisher must be installed on the forklift and shall be maintained in a serviceable condition.

23. The operating area shall be kept free of water, snow, ice, oil, and debris that could cause the operator's hands and feet to slip from the controls.

Picking Up and Moving Loads

It is important to know how much a load weighs before trying to move it. If the weight of the load is not clearly marked, try a simple test to see if it is safe to move. Lift the load an inch or two. Powered industrial trucks should feel stable and the rear wheels should be in firm contact with the floor. If everything is operating properly and steering seems normal begin to move the load. If the forklift struggles, set the load down and check with the supervisor before continuing. Operators need to practice picking up loads in various locations and in whatever situation they are expected to work.

All loads should be squared up on the center of the load and approached it straight on with forks in traveling position. Stop when the tips of the forks are about a foot away from the load. Level the forks and slowly drive forward until the load is resting against the backrest. Lift the load high enough to clear whatever is under it. Look in all directions to make sure the travel path is clear, and back out. Carefully tilt the mast back to stabilize the load.

Traveling with a Load

The nature of the terrain, the surface upon which the truck is to operate, is a very important factor in the stability of load-truck system. The designated person shall insure that a proper truck has been selected to operate on the surface available. In general, small, three-wheeled trucks are to be operated on smooth, hard surfaces only and are not suitable for outdoor work. The operator shall insure that the load is well secured and properly balanced before it is lifted. The lift is must be done slowly, with no sudden acceleration of the load, nor should it contact any obstruction. Here are some requirements for traveling in powered industrial trucks. Some of these requirements include:

1. All traffic regulations must be observed, including authorized plant speed limits.

2. The driver shall be required to slow down and sound the horn at cross aisles and other locations where vision is obstructed. If the load being carried obstructs the forward view, the driver is required to travel with the load trailing.

3. Railroad tracks shall be crossed diagonally whenever possible. Parking closer than eight feet from the center of railroad tracks is prohibited.

4. When ascending or descending grades in excess of ten percent, loaded trucks shall be driven with the load upgrade.

5. Always travel with a load tilted slightly back for added stability.

6. Travel with the load at the proper height. A stable clearance height is four to six inches at the tips and two inches at the heels to clear most uneven surfaces and avoid debris. (See Figure 12-4.)

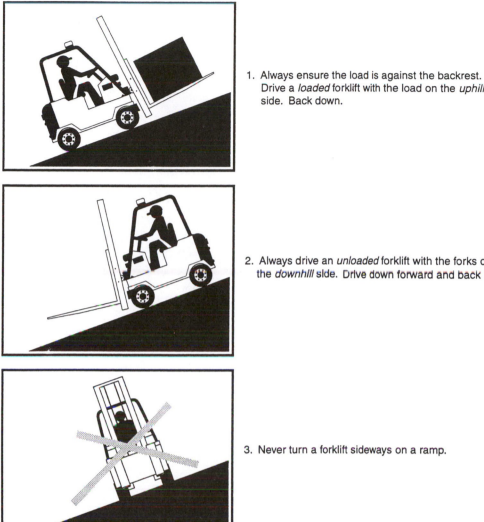

1. Always ensure the load is against the backrest. Drive a *loaded* forklift with the load on the *uphill* side. Back down.

2. Always drive an *unloaded* forklift with the forks on the *downhill* side. Drive down forward and back up.

3. Never turn a forklift sideways on a ramp.

Figure 12-4. Safe traveling for forklift trucks. Courtesy of the U.S. Department of Energy

1. Dockboards and bridgeplates shall be properly secured before they are driven over.

2. Dockboards and bridgeplates shall be driven over carefully and slowly and their rated capacity never exceeded.

3. Turning a powered industrial truck will require a little more concentration than driving a car. Because it steers from the rear, the forklift handles very differently from a car and other roadway vehicles. The back end of the forklift swings wide and can injure co-workers and damage products or equipment.

4. Once the load has been picked up, never make a turn at normal speed. Always slow down to maintain balance.

Stacking and Unstacking Loads

The use of powered industrial trucks to stack products and increase storage capacity is frequently undertaken. When stacking or unstacking a product, keep in mind that the higher the load is positioned, the less stable the truck becomes. Lifting a load from a stack is similar to lifting a load from the floor:

1. Approach the load slowly and squarely with the forks in the traveling position.

2. Stop about a foot from the load and raise the mast so the forks are at the correct height.

3. Level the forks and drive forward until the load is flush against the backrest.

4. Lift it high enough to clear the bottom load, look in all directions, and slowly back straight out.

5. Once the top of the stack has been cleared, stop and lower the mast to the traveling position. Tilt the forks back and proceed.

6. To stack one load on top of another, stop about a foot away from the loading area and lift the fork tips enough to clear the top of the stack.

7. Slowly move forward until the load is square over the top.

8. Level the forks and lower the mast until the load is no longer supported by the fork

9. Look over both shoulders and slowly back straight out.

10. Never lift a load while moving.

Standard Signals

Standard hand signals for use shall be as specified in the latest edition of the ANSI standards regarding powered industrial trucks. (See Figure 12-5.) The operator shall recognize signals from the designated signaler with the only exception being a STOP signal, which shall be obeyed no matter who gives it.

Safety Guards

All high-lift rider trucks shall be fitted with overhead guards, where overhead lifting is performed, unless operating conditions do not permit. In such cases where high-lift rider trucks must enter, as with truck trailers, when the overhead guard will not permit their entry, the guard may be removed, or a powered industrial truck without a guard may be used.

If a powered industrial fork truck carries a load that presents a hazard of falling back onto the operator, it shall be equipped with a vertical load backrest extension.

Trucks and Railroad Cars

In plant receiving and shipping areas, powered industrial trucks are often utilized to load and unload materials from trucks and railroad cars. The brakes of highway trucks shall be set and wheel chocks placed under the rear wheels to prevent trucks from rolling while they are boarded with powered industrial trucks.

Wheel stops or other positive protection shall be provided to prevent railroad cars from moving during loading or unloading operations.

RAISE THE TINES   LOWER THE TINES   TILT MASK BACK

TILT MASK FORWARD   MOVE TINES IN DIRECTION FINGER POINTS   DOG EVERYTHING

STOP

Figure 12-5. Hand signals for industrial (forklift) truck operation. Permission by the American Society of Mechanical Engineers

Fixed jacks may be necessary to support a semitrailer and prevent unending during the loading or unloading when the trailer is not coupled to a tractor.

Co-workers Safety

Never carry hitchhikers — they can easily fall off and become injured. If co-workers are on a safety platform, always ensure that the platform is securely attached to the forklift and personnel are wearing proper personal protective equipment (e.g. hardhats and safety harness). Never travel with co-workers on the platform. Watch out for overhead obstructions.

Pedestrian Safety

Pedestrians working nearby should be sure to keep a safe distance from forklifts. That means staying clear of the forklift's turning radius and making sure the driver knows where you are.

Conduct of the Operator

The operator's driving skill, attitude, adherence to safety rules, and conduct will play an important role in powered industrial truck safety. The operator should:

1. Not engage in any practice which will divert attention while operating the powered industrial truck.

2. Not operate the forklift when physically or mentally incapacitated.

3. Before operation of electric powered machines, check location of the battery plug for quick disconnection in case of a short circuit.

4. Avoid sudden stops.

5. Face in the direction of travel, except as follows:

    a. For better vision with large loads, operate the truck in reverse gear.

    b. Do not descend ramps with the load in front.

6. Watch blind corners, stop at all intersections and doorways, and sound the horn.

7. Operate at safe speeds: in-plant buildings—5 miles per hour; on-plant roads–15 miles per hour maximum.

8. Go slow around curves.

9. Use low gear for the slowest speed control when descending ramps.

10. Know the rated capacity of the truck and stay within it.

11. Consider both truck and load weight.

12. Watch overhead clearance. If in doubt, measure.

13. Keep clear of the edge of the loading dock.

14. Watch rear-end swing.

15. Before handling, assure that stacks and loads are stable. Block and lash them if necessary.

16. Always spread the forks to suit the load width.

17. Lower and raise the load slowly. Make smooth gradual stops.

18. Lift and lower loads only while the vehicle is stopped.

19. Use special care when high-tiering. Return the lift to a vertical position before lowering load.

20. Lift, lower, and carry loads with the upright vertical tilted back, never forward.

21. To avoid personal injury, keep arms and legs inside the operator's area of the machine.

22. Never travel with forks raised to unnecessary heights. Approximately four to six inches above floor level is adequate.

23. When loading trucks or trailers, see that the wheels are chocked and the brakes set. Operate in front end of the semitrailer only if the tractor is attached, or adequate trailer (railroad) jacks are in place.

24. Inspect floors on trucks, boxcars, unfamiliar ramps, or platforms before start of operation.

25. Be sure bridge plates into trucks or freight cars are sufficiently wide, strong, and secure.

26. Never butt loads with forks or rear end of truck.

27. Fork trucks should not be used as tow trucks. They are built for lifting only, unless a towing hitch is supplied by the manufacturer. Use tow bars rather than cable for towing.

28. Stop engine before refueling.

29. Use only approved explosion-proof lights to check gas tank and battery water levels.

30. Smoking is not permitted during this operation.

31. Place forks flat on the floor when truck is parked.

32. Turn switch key off when leaving the machine.

33. Always set brakes before leaving the truck.

34. Report evidence of faulty truck performance.

35. When alighting from truck, step down — do not jump.

36. Report all accidents promptly to your supervisor.

Operators who are properly trained are expected to adhere to all of the previously iterated requirements for operator conduct and safe work practices when using powered industrial trucks.

TRAINING OF OPERATORS

As of March 1, 1999, employers who use powered industrial trucks (forklifts) in the general industry, construction, or maritime industries must comply with OSHA's new forklift training standards, 29 CFR 1910.178(l), 29 CFR 1915.120 and new 1926.602(d), which are identical to 1910.178(l) and CFR 1917 and 1918, which includes the training requirements by reference to 178(l). About 100 workers are killed each year in incidents related to industrial truck operation and nearly 95 thousand suffer injuries each year that result in lost workdays. Approximately 30 percent of these incidents are at least in part caused by inadequate training. (See Figure 12-6 and 1-7)

Federal regulations on training of all forklift operators are in 29 CFR 1910.178. These regulations require that only trained and authorized persons shall be permitted to operate a powered industrial truck, the regulatory definition of "forklift." This includes all employees who may use a forklift, even if it is only a casual or occasional part of their job duties.

Figure 12-6. Example of powered industrial truck

Figure 12-7. Example of a forklift

General Requirements

The employer must ensure that each powered industrial truck operator is competent to operate a powered industrial truck safely, as demonstrated by the successful completion of training and evaluation. Prior to permitting an employee to operate a powered industrial truck (except for training purposes), the employer shall ensure that each operator has successfully completed the required training.

While implementing training, trainees may operate a powered industrial truck under the direct supervision of persons who have the knowledge, training, and experience to train operators and evaluate their competency, and where such operation does not endanger the trainee or other employees.

All training is to consist of a combination of formal instruction (e.g., lecture, discussion, interactive computer learning, video tape, or written material), practical training (demonstrations performed by the trainer and practical exercises performed by the trainee), and evaluation of the operator's performance in the workplace.

The employer shall insure that all operator training and evaluation shall be conducted by persons who have the knowledge, training, and experience to train powered industrial truck operators and evaluate their competence.

Training Program Content

Powered industrial truck operators must receive initial training in the following topics, with the exception of topics that the employer can demonstrate are not applicable to safe operation of the truck in his or her workplace. The topics are as follows:

- Operating instructions, warnings, and precautions for the types of truck the operator will be authorized to operate.
- Differences between the truck and the automobile.
- Truck controls and instrumentation: where they are located, what they do, and how they work.
- Engine or motor operation.
- Steering and maneuvering.
- Visibility (including restrictions due to loading).
- Fork and attachment adaptation, operation, and use limitations.
- Vehicle capacity.
- Vehicle stability.
- Any vehicle inspection and maintenance that the operator will be required to perform.
- Refueling and/or charging and recharging of batteries.
- Operating limitations.
- Any other operating instructions, warnings, or precautions listed in the operators' manual for the types of vehicle that the employee is being trained to operate.

The training must also consist of specific workplaces related hazards and topic such as:

- Surface conditions where the vehicle will be operated.
- Composition of loads to be carried and load stability.

- Load manipulation, stacking, and unstacking.
- Pedestrian traffic in areas where the vehicle will be operated.
- Narrow aisles and other restricted places where the vehicle will be operated.
- Hazardous (classified) locations where the vehicle will be operated.
- Ramps and other sloped surfaces that could affect the vehicle's stability.
- Closed environments and other areas where insufficient ventilation or poor vehicle maintenance could cause a buildup of carbon monoxide or diesel exhaust.
- Other unique or potentially hazardous environmental conditions in the workplace that could affect safe operation.

Refresher Training and Evaluation

Refresher training, including an evaluation of the effectiveness of that training, shall be conducted to ensure that the operator has the knowledge and skills needed to operate the powered industrial truck safely. Refresher training in relevant topics shall be provided to the operator when:

- The operator has been observed to operate the vehicle in an unsafe manner.
- The operator has been involved in an accident or near-miss incident.
- The operator has received an evaluation that reveals that the operator is not operating the truck safely.
- The operator is assigned to drive a different type of truck.
- A condition in the workplace changes in a manner that could affect safe operation of the truck.

Reevaluation

An evaluation of each powered industrial truck operator's performance shall be conducted at least once every three years.

Avoidance of Duplicative Training

If an operator has prior training in the previously specified topics, and such training is appropriate to the truck and present working conditions encountered, and the operator has been evaluated and found competent to operate the truck safely, then additional training in that topic is not required.

Certification

The employer shall certify that each operator has been trained and evaluated as required. Certification shall include the name of the operator, the date of the training, the date of the evaluation, and the identity of the person(s) performing the training and evaluation.

In-House Training Development

Training programs should be tailored to employees' work situations. Employees benefit more from training that simulates their daily processes, rather than from watching "canned" programs that are not applicable to their specific operations. Training programs should be devised so that employees can demonstrate the knowledge and skills required for their job.

Driving Skill Evaluations

A key dimension of operator training is driver certification. Operators should be required to demonstrate their skills. Adequate completion of skills tests (recorded on rating sheets similar to those in Tables 12–3, 12–4, and 12–5) demonstrates both that the operator knows and understands the unit's functional features, and is familiar with overall departmental safety rules and can identify specific safety factors at a dock and battery recharge station. She/he must also demonstrate overall driving skills. Testing can be administered on the job during the employee's normal workday.

Table 12-3

Driving Skills Evaluation Checklist
Courtesy of U.S. Department of Energy

The objective of this rating sheet is to ensure that employees understand the mechanics of the lift truck and of those items involving standard checking prior to driving the lift truck. The operator should be familiar with the features of the specific type of lift truck h/she is to operate. This can be evaluated by having the operator demonstrate and describe the following:

_____ Proper use of tilt		_____ Check scissors reach.
_____ Proper use of raise and lower.		_____ Check warning light.
_____ Proper use of horn.		_____ Check rear view mirror.
_____ Check for oil leaks.		_____ Check battery retainer.
_____ Check mast chains.		_____ Check discharge indicator.
_____ Check tilt and lift cylinders for wear and/or leakage.		_____ Check back up alarm.
		_____ Check hose and hose reel.
_____ Check brakes.		_____ Check over head guard's light.
_____ Check tires and wheels.		_____ Know capacity of lift truck.
_____ Check hour meter.		

Table 12-4

Knowledge of Safeguards within the Facility Checklist
Courtesy of U.S. Department of Energy

The operator is asked to identify many safety items at the dock and battery recharging area, as well as overall safety.

Dock	**Battery Charging Area**
_____ Wheel chocking.	_____ Protective equipment.
_____ Dock plate. _____ Acid neutralizing.	
_____ Trailer lighting	_____ MSDS.
_____ Condition of trailer floor.	_____ No smoking.
_____ Keep clear of dock loading area.	_____ Plug/unplug procedures.
_____ Be aware of signs. _____ Clean-up procedures.	
_____ Correct height of empty pallets.	_____ Eyewash station.
_____ Commercial battery rules.	

Fire and Safety **Personal Safety**	
_____ Location of extinguishers.	_____ Use of eye protection during banding operations.
_____ How to use extinguisher.	
_____ Type of extinguisher to use.	
_____ Eye protection during banding.	

Table 12-5

Operating Skills Evaluation Checklist
Courtesy of U.S. Department of Energy

Determine the operating skills of employees by making a full evaluation while they are driving the lift truck. The following should be checked:

_____ Did the operator pull forward toward the designated section of racking without endangering anyone?

_____ Did the operator place the forks under the pallet properly?

_____ Did the operator raise or tilt the load properly?

_____ Did any part of the container strike any section of racking while removing the pallet?

_____ Did the operator lower the pallet before moving or backing out? (Don't drive and lower the pallets at the same time.)

_____ Did the operator drive at a safe speed?

_____ Did the operator slow down or stop at cross aisles?

_____ Did the operator sound his/her horn at blind intersections?

_____ Did the operator pull into the racking area properly to place the pallet back in the racking?

_____ Did the operator strike any racking on the way up or going into the rack?

_____ Did the operator back out and lower his/her forks before moving?

_____ Did the operator always look behind before backing up?

_____ Was the operator wearing protective equipment?

_____ Did the operator drive around the block of wood or obstacle on the floor, or did he/she get out of the truck and remove it?

_____ Did the operator set the load flat on the floor before getting out of the truck?

_____ Did the operator put on a hardhat before getting out of the truck?

_____ Did the operator perform any moves that were potentially dangerous?

STANDARDS AND REGULATIONS

The following is a list of the applicable standards relevant to powered industrial trucks from varied official organizations.

Organization	Standard	Title
OSHA	29 CFR 1910.178	Powered industrial trucks.
OSHA	29 CFR 1910.1000	Air contaminants.
OSHA	29 CFR 1926.602	Material handling equipment.
ANSI	B56.1—1988	American national standard for powered industrial trucks.
NFPA	NFPA No. 30—1969	NFPA flammable and combustible liquids code.
NFPA	NFPA No. 58—1969	NFPA storage and handling of liquefied petroleum gases.
NFPA	NFPA No. 505—1969	Powered industrial trucks.
UL	583	Standard for safety for electric or battery-powered industrial trucks.
UL	558	Standard for safety for internal com bustion or engine-powered industrial trucks.
ANSI/NFPA	30—1987	Flammable and combustible liquid code.

| ANSI/NFPA | 58–1986 | Storage and handling of liquefied petroleum gases. |
| ANSI/NFPA | 505–1987 | Fire safety standard for powered industrial trucks—type designations, areas of use, maintenance and operation. |

OSHA = Occupational Safety and Health Administration
ANSI = American National Standards Institute
NFPA = National Fire Protection Association
UL = Underwriters Laboratory

REFERENCES

American National Standards Institute. *Combustion Engine-Powered Industrial Trucks*. No. 558 (ANSI B56.4–1980). New York, NY, 1980.

"Forklift Training." *Industrial Maintenance and Plant Operation*. p. 34. January 1990.

"Lift Truck Training: It's Here and It Works." *Modern Materials Handling,* pp. 72-78. September 1989.

"Operator Training: More than Just Driving a Lift Truck." *Materials Handling Engineering*. pp. 33-46. June 1990.

Moran, Mark McGuire. *Construction Safety Handbook*. Government Institutes Inc. Rockville, MD, 1996

Reese, C. D. and J.V. Eidson. *Handbook of OSHA Construction Safety & Health*. Boca Raton, FL: CRC/Lewis Publishers, 1999.

United States Department of Energy. *Hoisting and Rigging Manual*. Washington, DC, 1991.

United States Department of Energy. *OSH Technical Reference Manual*. Washington, DC, 1993.

United States Department of Labor. Occupational Safety and Health Administration. *OSHA 10 and 30 Hour Construction Safety and Health Outreach Training Manual*. Washington, DC, 1991.

United States Department of Labor. Occupational Safety and Health Administration. General Industry. *Code of Federal Regulations*. Title 29, Part 1910. Washington, GPO, 1998.

United States Department of Labor. Occupational Safety and Health Administration. Construction. *Code of Federal Regulations*. Title 29, Part 1926. Washington, GPO, 1998.

United States Department of Labor. Occupational Safety and Health Administration. Office of Training and Education. *OSHA Voluntary Compliance Outreach Program: Instructors Reference Manual*. Des Plaines, IL, 1993.

CHAPTER 13

RAILROAD CARS

Typical railroad yard

In a discussion of material handling relevant to railroad cars, just the size and volume of material which they can hold presents a unique problem. Beyond this, railroad cars transport a diverse variety of materials ranging from livestock to dangerous chemicals.

The movement and handling of railroad cars throughout history has been an extremely dangerous task. The size and weight of these types of material handling and transporting devices make them a unique hazard each time that they moved. Suffice it to say a loaded railroad car in motion has enormous inertial energy, which means stopping them is a difficult task at best. But with the invention of air brakes for railroad cars, this task became easier and safer. The air brakes lessened the hazard of being run over or crushed by a railroad car. Also, the invention of the automatic coupler provides a safety feature when coupling and uncoupling railroad cars. The large numbers of amputation of hands and fingers, which was very prevalent prior to this invention, has significantly decreased. (See Figure 13-1.)

A pair of gloves for a
CARELESS
car coupler

Figure 13-1. A pair of gloves for workers whose job it was to couple and uncouple railroad cars.

Railroad cars may be freight cars, tanker cars, bulk material cars, or those designed for special hauling purposes. Freight cars may be used to contain anything from palletized material to livestock. Tanker cars usually contain various liquids and often concentrated hazardous chemicals, which pose unique handling and transportation problems. Bulk material cars may be completely enclosed for such materials as grain and open for hauling coal or other mining products. Specially designed railroad cars are those used to haul such items as new automobiles and large machinery.

The size, weight, and narrow range of movement (forward and backward) are issues faced when attempting to safely handle materials by railroad cars. The loading or off-loading process usually requires special docking or handling facilities, since railroad cars can only be maneuvered in two directions. In most cases the movement of railroad cars depends on the presence of a yard engine (See Figure 13-2) being available to facilitate the positioning and movement of the railroad cars. Whether cars are loaded or unloaded, the number of cars and the lay of the tracks, whether flat or on a grade, must be taken into consideration when using them for material handling.

When power equipment is not used to move railroad cars, a car mover can be used to manually move a car into position. This activity is best done with two workers coordinating the move. The car mover should have knuckle guards for safety. Once cars come to a rest and are ready for loading and unloading, the brakes must be set and wheels should be chocked to prevent errant or unexpected movement of the railroad cars.

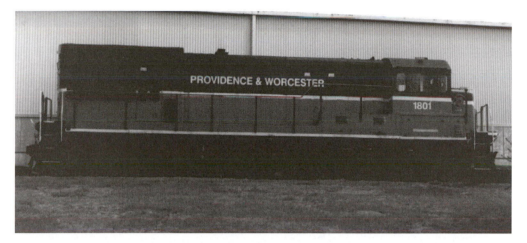

Figure 13-2. An example of a typical yard engine used for moving railroad cars

FREIGHT RAIL CARS

Freight rail cars are similar to the trailers which are used in the long haul trucking industry. (See Figure 13-3.) Thus, although not as maneuverable as long haul trailers, they can be loaded or unloaded with the same materials, just more and heavier, using docking facilities, secured dockplates, forklifts or by manual means. Also, ramps can be used, if secured, when no formal docking facilities exist. Normally, the safe operating procedures for the type of material being transported and handled would be followed.

Figure 13-3. A typical freight rail car

During transport, materials may shift. Workers should not only be aware of getting their fingers pinched when opening the doors, they should be also alert to potential hazards from falling materials when doors are opened. The possibility of falling freight or materials is always a hazard faced by workers.

TANKER RAIL CARS

Railroad tanker cars usually present unique problems, since their contents may be hazardous or toxic and there always exists the potential for spills. Liquids are of special concern because flowing makes containment difficult. (See Figure 13-4.) Tanker cars require specially designed loading and unloading equipment, since materials must be pumped and specific procedures may be required. When working around railroad tanker cars, workers must be alert for leaking materials. If leaks are detected, emergency response procedures may need to be instituted. Also, special care must be taken to prevent damaging valves or piping on the tanker cars.

Figure 13-4. An example of a railroad tanker car

No worker should ever enter a tanker car for cleaning or purging without following confined space entry procedures where oxygen deficiency, flammable, or toxic atmospheres may exist. Close attention should be paid to placards or warning signs on tanker cars.

BULK MATERIAL RAIL CARS

Bulk material railroad cars transport materials, which, usually, are loaded from overhead hoppers, bins, conveyors, silos, augers, shovels, frontend loaders, or by other means. Typically these types of railroad cars are off-loaded by opening hatches on the bottom to the cars. (See Figure 13-5.) During off-loading, materials may bridge, creating a void beneath materials, or the same materials may adhere to the sides of the bulk rail cars. This may be due to weather conditions (like freezing) or moisture content affecting the bulk material. If a worker enters the car and steps on the bridged area, he or she may collapse the bridge and be engulfed. At times workers enter bulk cars to dislodge materials, which also might result in engulfment. When workers must enter bulk cars, they should wear a harness and lifeline. Care must be taken to have a lifeline that does not allow much freedom of movement, or it may be of no protection if the worker is engulfed by bulk materials. The approximate weight of a cubic yard of bulk material is 3,000 pounds. A worker could not be pulled free by using the lifeline if he or she is engulfed. Thinking before acting, in this regard, can save lives.

Figure 13-5. A bulk material rail car

FLATBED RAIL CARS

Flatbed rail cars provide flexibility for the handling and hauling of a variety of oddly shaped equipment, containers, and materials. Flatbed rail cars are usually reasonably low with few obstructions. This makes the loading and unloading less complicated. Their openness allows for a greater variety of uses than any other type of rail car. (See Figure 13-6.) Insuring that loads are balanced and secured provides a higher level of safety and compliance with regulations.

Figure 13-6. A flatbed rail car

HOPPER RAIL CARS

At times, hauling material which does not need to be covered or protected from the weather is accomplished by the using hopper rail cars. Some of these cars are very deep; others are shallow. Which is used depends upon the product being hauled. These rail cars have gates on the bottom, which can be opened to expel materials carried within. Hopper car gates are very stubborn at times and require special tools to open them. Both manual and powered tools are available to assist in unloading hopper cars. (See Figure 13-7 and 13-8.) Care must be taken when loading, unloading, or entering hopper cars due to shifting of the materials being handle.

Figure 13-7. A hopper rail car

Figure 13-8. Opening gates on hopper rail cars. Permission by the Aldon Co.

SPECIALLY DESIGNED RAIL CARS

Specially designed rail cars for specific purposes are not commonplace but do find their place in the railroad industry. These cars may serve the purpose of carrying a variety of items form new automobiles to carrying livestock. The need, of course, determines the specification and design of specialty rail cars. (See Figure 13-9.) Specially designed rail cars may have unique safety requirements, which should be conveyed clearly to all workers involved with their use by manufacturers, employers, supervisors, and co-workers.

Figure 13-9. Specially designed rail cars

TOOLS AND EQUIPMENT FOR RAIL CAR HANDLING

The railroad industry has been the impetus for the development of many tools and types of equipment to assure the safe handling of railroad cars as well as the safe loading and unloading of them. Most tools and equipment for this purpose are manually powered, but some inroads have been made to make them power-assisted.

Most of the equipment can be classified as chocks, car stopping skids, bump posts, derailers, car pullers, car movers, hopper car gate openers, boxcar door openers, dockboards, loading ramps, hopper shakers, or coupler alignment tools.

Chocks are used to assure that rail cars do not move during loading and unloading. Car stoppers are used to keep the car from passing a certain point. (See Figure 13-10.) Derailers are put in place to stop a runaway rail car. They cause the wheels to be kicked off the track and, usually, bring the car to a stop. (See Figure 13-11.) When dealing with rail cars, errant movement is always a possibility, so the use of safety devices to prevent or stop movement is imperative to worker safety.

Figure 13-10. Chocks and car stopping skids. Permission by The Aldon Co.

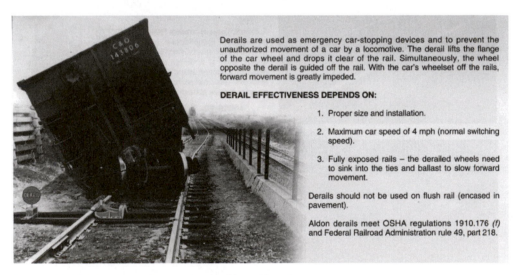

Derails are used as emergency car-stopping devices and to prevent the unauthorized movement of a car by a locomotive. The derail lifts the flange of the car wheel and drops it clear of the rail. Simultaneously, the wheel opposite the derail is guided off the rail. With the car's wheelset off the rails, forward movement is greatly impeded.

DERAIL EFFECTIVENESS DEPENDS ON:

1. Proper size and installation.

2. Maximum car speed of 4 mph (normal switching speed).

3. Fully exposed rails – the derailed wheels need to sink into the ties and ballast to slow forward movement.

Derails should not be used on flush rail (encased in pavement).

Aldon derails meet OSHA regulations 1910.176 *(f)* and Federal Railroad Administration rule 49, part 218.

Figure 13-11. Derailers. Permission by The Aldon Co.

Car pullers allow a worker to actually pull a rail car into position for loading and unloading. Car pullers are devices, usually run by electricity, which provide a mechanical advantage to the worker trying to move the car. (See Figure 13 -12.) A car mover can be a manual leverage device, or powered by a gasoline engine, which allows a worker to move a stationary car into position for loading or unloading. (See Figure 13-13.) Workers need to be trained to use these devices in order to prevent injuries and equipment damage from occurring.

Figure 13-12. Car pullers. Permission by The Aldon Co.

Figure 13-13. Car movers. Permission by The Aldon Co.

Other equipment used to load and unload rail cars comes in all types of designs from simple manual devices to powered equipment, which negates most of the physical force that a worker would have to use. As stated earlier, any time an effort can be mechanized, the chance of injury is greatly reduced.

Railroad cars are important vehicle in the handling of material in large quantities. Their use can be carried out in a safe manner without the loss or life or limbs, or other injuries and illnesses.

REFERENCES

United States Department of Labor. Occupational Safety and Health Administration. General Industry. *Code of Federal Regulations*. Title 29, Part 1910. Washington, GPO, 1998.

United States Department of Labor. Occupational Safety and Health Administration. Office of Training and Education. *OSHA Voluntary Compliance Outreach Program: Instructors Reference Manual.* Des Plaines, IL, 1993.

CHAPTER 14

INDUSTRIAL ROBOTS

Industrial robot with barrier

Industrial robots are used within the workplace for a wide variety of tasks. One task is material handling. Robots have what might be called a mind of their own. Once programmed and set into action, they go about their tasks. Industrial robots complete their tasks rapidly, efficiently, forcefully, and without ceasing. The robot is not going to be able to sense the presence of a worker within its path of movement, and certainly will not stop when a worker is in the way. Thus, robots, which operate with great force and speed, are dangerous to workers, and precautions and training regarding industrial robots must be undertaken.

213

Industrial robots are programmable, multifunctional, mechanical devices designed to move material, parts, tools, or specialized devices through variable programmed motions to perform a variety of tasks. An industrial robot system includes not only industrial robots, but also any devices and/or sensors required for the robot to perform its tasks as well as sequencing or monitoring communication interfaces.

Robots are generally used to perform unsafe, hazardous, highly repetitive, and unpleasant tasks. They have many different functions such as material handling, assembly, arc welding, resistance welding, machine tool load and unload functions, painting, spraying, etc. Most robots are set up for an operation by the teach-and-repeat technique. In this mode, a trained operator (programmer) typically uses a portable control device (a teach pendant) to teach a robot its task manually. Robot speeds during these programming sessions are slow. Instructions for trained operators usually include safety considerations necessary to operate the robot properly and use it automatically in conjunction with other peripheral equipment.

The Occupational Safety and Health Administration (OSHA) has not undertaken the development of a regulation specifically for robots. Chapter four of *OSHA's Technical Manual* addresses guidelines for industrial robots and robot system safety; these guidelines are incorporated into the content of this chapter. This chapter includes information on how to properly and safely operate fixed industrial robots and robot systems with other peripheral equipment.

INCIDENCE OF ROBOTIC ACCIDENTS

Studies in Sweden and Japan indicate that many robot accidents occur during programming, program touchup, maintenance, repair, testing, setup, or adjustment rather than under normal operating conditions. The operator, programmer, or corrective-maintenance worker may temporarily be within the robot's working envelope where unintended operations can result in injury. For example:

- A robot's arm functioned erratically during a programming sequence and struck the operator.

- A materials-handling-robot operator entered a robot's work envelope during operations and was pinned between the back end of the robot and a safety pole. (See Figure 14 -1.)

- A fellow employee accidentally tripped the power switch while a maintenance worker was servicing an assembly robot. The robot's arm struck the maintenance worker's hand.

Causes of Robotic Accidents

Some of the unsafe condition and acts which have contributed to robot-related incidents in the past are:

Unsafe Acts:

- Placing oneself in hazardous positions while programming or performing maintenance within the robot's work envelope.

- Inadvertently entering the envelope because of unfamiliarity with the safeguards in place or not knowing if they are activated.

- Making errors in programming, interfacing peripheral equipment, and connecting input/output sensors.

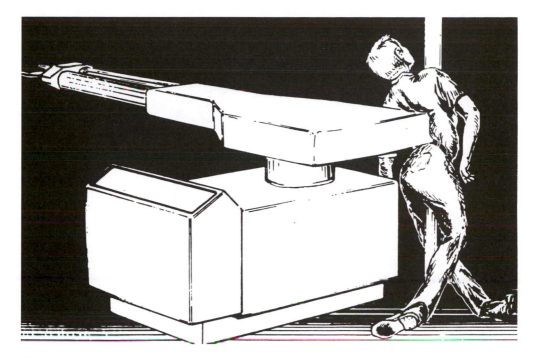

Figure 14-1. Robot-related fatality. Courtesy of the National Institute for Occupational Safety and Health

Unsafe Conditions:

- Mechanical failure.

- Safeguards deactivated.

- Intrinsic faults within the control system of the robot, errors in software, electromagnetic interference, and radio-frequency interference.

- Hazards from pneumatic, hydraulic, or electrical power that can result from malfunction of control or transmission elements of the robot's power system such as control valves, voltage variations, or voltage transients disrupting the electrical signals to the control and/or power supply lines.

- High temperature ignition that can result from electrical overloads or from the use of flammable hydraulic oil.

- Electrical shock and release of stored energy from accumulating devices that can result in injury to personnel.

HAZARDS

The operational characteristics of robots can be significantly different from other machines and equipment. Robots are capable of high-energy (fast or powerful) movements through a large volume of space even beyond the base dimensions of the robot. The pattern and initiation of movement of the robot is predictable if the item being "worked" and the environment are held constant. Any change to the object being worked (i.e., a physical model change) or the environment can affect the programmed movements.

Some maintenance and programming personnel may be required to be within the restricted envelope while power is available to actuators. The restricted envelope of the robot can overlap a portion of the restricted envelope of other robots or work zones of other industrial machines and related equipment. Thus, a worker can be hit by one robot while working on another, trapped between them or peripheral equipment, or hit by flying objects released by the gripper.

A robot with two or more resident programs can find the current operating program erroneous, therefore calling another existing program with different operating parameters such as velocity, acceleration, or deceleration, or position within the robot's restricted envelope. The occurrence of this might not be predictable by maintenance or programming personnel working with the robot. A component malfunction could also cause an unpredictable movement and/or robot arm velocity.

Additional hazards can also result from the malfunction of, or errors in, interfacing or programming of other processes or peripheral equipment. The operating changes with the process being performed or the breakdown of conveyors, clamping mechanisms, or process sensors could cause the robot to react in a different manner.

Types of Hazards

Robotic incidents can be grouped into four categories: a robotic arm or controlled tool causes the accident or places an individual in a risk circumstance, an accessory of the robot's mechanical parts fails, or the power supplies to the robot are uncontrolled.

- Impact or collision accidents—Unpredicted movements, component malfunctions, or unpredicted program changes related to the robot's arm or peripheral equipment can result in contact accidents.

- Crushing and trapping accidents—A worker's limb or other body part can be trapped between a robot's arm and other peripheral equipment, or the individual may be physically driven into and crushed by other peripheral equipment.

- Mechanical part accidents—The breakdown of the robot's drive components, tooling or endeffector, peripheral equipment, or its power source is a mechanical accident. The release of parts, failure of gripper mechanism, or the failure of end-effector power tools (e.g., grinding wheels, buffing wheels, deburring tools, power screwdrivers, and nut runners) are a few types of mechanical failures.

- Other Accidents — Equipment that supplies robot power and control represents potential electrical and pressurized fluid hazards. Ruptured hydraulic lines could create dangerous high-pressure cutting streams or whipping hose hazards. Environmental accidents from arc flash, metal spatter, dust, electromagnetic, or radio-frequency interference can also occur. In addition, equipment and power cables on the floor present tripping hazards.

Sources of Hazards

Hazards inflicted by machines to humans can be expected with several additional variations, as follows:

- Human errors—Inherent prior programming, interfacing activated peripheral equipment, or connecting live input-output sensors to a microprocessor or a peripheral can cause dangerous, unpredicted movement or action by the robot from human error. The incorrect activation of the "teach pendant" or control panel is a frequent human error. The greatest problem, however, is overfamiliarity with the robot's redundant motions so that an individual places himself in a hazardous position while programming the robot or performing maintenance on it.

- Control errors—Intrinsic faults within the control system of the robot, errors in software, electromagnetic interference, and radio frequency interference are control errors. In addition, these errors can occur due to faults in the hydraulic, pneumatic, or electrical subcontrols associated with the robot or robot system.

- Unauthorized access—Entry into a robot's safeguarded area is hazardous because the person involved may not be familiar with the safeguards in place or their activation status.

- Mechanical failures—Operating programs may not account for cumulative mechanical part failure, and faulty or unexpected operation may occur.

- Environmental sources—Electromagnetic or radio-frequency interference (transient signals) can exert an undesirable influence on robotic operation and increase the potential for injury to any person working in the area. Solutions to environmental hazards should be documented prior to equipment start-up.

- Power systems—Pneumatic, hydraulic, or electrical power sources that have malfunctioning control or transmission elements in the robot power system can disrupt electrical signals to the control and/or power-supply lines. Fire risks are increased by electrical overloads or by use of flammable hydraulic oil. Electrical shock and release of stored energy from accumulating devices also can be hazardous to personnel.

- Improper installation—The design, requirements, and layout of equipment, utilities, and facilities of a robot or robot system, if inadequately done, can lead to inherent hazards.

Prevention Overview

System components must be designed, installed, and secured so that the hazards associated with stored energy are minimized. Adequate room must be provided for a robot's movement as well as for workers. There must be a means for controlling the release of stored energy in all the robotic systems and for shutting off power from outside the restricted envelope. A detailed risk assessment should be performed to ensure the safety of workers who operate, service and maintain the robotic systems.

Risk Assessment

A risk assessment (also termed "hazard analysis") should be required for robotic installations. Risk assessments are part of a safety analysis. Documentation preparation and updating technical safety requirements and procedures when robotic installations have safety hazard implications should be part of a good risk assessment and communications program for safety's sake.

Risk assessments for robot installations are based on the satisfaction of current criteria, regardless of the age of the basic facility or the date of the design of the equipment to be installed. Although closely linked to system design and preparation for use, risk assessments must be done independently using a systematic approach to identifying and assessing hazards. Designers and users should be involved in identifying approaches to reduce risks, but the assessment of residual risk must be performed independently and objectively. Note: "Current criteria" include all applicable federal regulations and the most recent standards, specifications, or criteria cited in other applicable standards. The criteria contained in guidance documents issued or cited by federal agencies or contained in regulations are usually relevant to specific facilities or operations.

Preparation of a preliminary risk assessment early in the conceptual design is desirable to meet safety objectives most effectively. Needed remediation and mitigation measures can determine the choices among significantly different concepts. These choices will also further affect design definition, layout, and component selection. Safety-related sensors, switches, and interlocks must be used where hazards are not reduced sufficiently by system and facility design. Safe operating procedures are required to complement passive and active safety systems.

The assessment of residual risk is based upon the sum of benefits provided by multiple approaches to meeting safety objectives. Management must review the risk assessments and safety analyses and approve the means of safeguarding. The following factors must be addressed in a risk assessment of robotic systems:

- The size, capacity, and speed of the robot.

- The specific application and associated processes.

- The anticipated tasks required for continued operation.

- The hazards associated with identified tasks, applications, and associated processes.

- Anticipated failure modes, including human errors and system malfunctions.

- The probability of occurrence of potential failures and estimated severity of injury associated with these anticipated failures.

- The capability of meeting the levels of expertise required for personnel. Other factors that may be appropriate to consider during a risk assessment are:

  - Adequacy of testing and start-up procedure.

  - Adequacy of training programs.

  - Satisfaction of installation criteria.

  - Environmental considerations.

  - Satisfaction of occupational safety and health criteria.

  - Maintenance and inspection activities.

  - The system stage(s) of development.

Safeguarding requirements vary for different stages of robot system development. The probability of occurrence of hazards can be dependent on factors such as numbers of people present, levels of experience, and types of robot application. Nonroutine use is typically more hazardous than routine use. Extended application engineering and setups involving multiple tools and robot system interfaces add to the probability of hazards.

The stages of development for a new system can include testing, research, and design up through and including productive use. Risk assessment shall recognize different situations and the unique safeguarding measures that shall be designed accordingly. For example, frequent access to the restricted envelope may be anticipated. If safety depends on a system of fixed barriers, the probability that barriers may not be replaced and that interlocks will be overridden must be considered. These may be considered by development personnel and others as unnecessarily restricting essential tasks. The safeguarding system, the training, procedures, and managerial oversight would all need to recognize the situation.

Risk assessment must address the current stages of system development and must be revised as system hazards and stages of development unfold. Features that give a modern industrial robot its value can also be principal sources of hazards. Robots achieve flexibility in application through the variety of motions which can be programmed, the ease with which the programming can be changed, the speed of movement, and the large production volume. These same features can cause human entry into the robot's envelope to be hazardous. In fact, human entry into this space has been the primary cause of robot-related deaths and injuries. Designing for safety, therefore, becomes a compromise between functional capabilities and hazard minimization.

Preventing Robotic Accidents

The majority of accidents involving robots could be prevented if robot installation layouts and procedures followed guidelines outlined in the robot safety standards by different agencies. Industrial robots are available (new, used, and rebuilt) that do not meet recommended industrial robot safety standards or installation requirements. Management is responsible for ensuring that robot systems are procured and installed to meet the following safety requirements:

- Robot systems must incorporate shielding, filtering, or suppression—This will prevent hazardous motion caused by radio frequency or electromagnetic interference and electrostatic discharge.

- These systems must allow each axis to be moved without using drive power—This feature permits correction of a pin or pinch accident after deenergizing the robot. This feature also allows for maintenance or testing of tasks without having to energize the robot where movement of the robot arm is required.

- The robot must have slow speed capabilities, where the maximum speed of any part of the robot is less than 150 millimeters (6 inches) per second—This speed has been determined appropriate to allow personnel to react in adequate time and avert the hazard. Having the capability of operating at this or a slower speed is mandatory for any robot where there is to be any human access to the restricted envelope with the robot energized.

- All incorporated and installed mechanical stops must be capable of stopping robot motion under rated load and maximum speed conditions—In the absence of adequate stops, the robot's maximum envelope must be used as the restricted envelope to prevent human access.

- Provisions for lifting the robot and associated equipment must be provided and must be adequate for handling the anticipated load.

- All electrical connectors that could cause a hazardous situation if disconnected must be specially marked, labeled, and guarded against unintended separation or mismating.

- Robots and their associated system components must be designed and constructed so that loss of electrical power, voltage surges, and changes in oil or air pressure do not result in hazardous motions. Any hoses incorporated in the system must be secured or protected to minimize hazards that may be caused by hose failure or disconnection.

TYPES AND CLASSIFICATION OF ROBOTS

Industrial robots are available commercially in a wide range of sizes, shapes, and configurations. They are designed and fabricated with unique system configurations, such as different numbers of axes or degrees of freedom, and they may handle a huge variety of different materials. These factors and others affect the robot's design, influence its working envelope, and the volume of working or reaching space. (Definition regarding robot nomenclature can be found in the glossary of this book.) Types of robots and their applications vary greatly.

All industrial robots are either classified servo or nonservo controlled. Servo robots are controlled through the use of sensors that continually monitor the robot's axes and associated components for position and velocity. This feedback is compared to pretaught information, which has been programmed and stored in the robot's memory. Nonservo robots do not have feedback capability, and their axes are controlled through a system of mechanical stops and limit switches. Both have safety issues inherent in their design.

Type of Path Generated

Industrial robots can be operated from a distance to perform their required and preprogrammed operations with different types of paths generated through different control techniques. The three different types of paths generated are point-to-point path, controlled path, and continuous path.

Point-to-Point Path

Robots programmed and controlled in this manner are programmed to move from one discrete point to another within the robot's working envelope. In the automatic mode of operation, the exact path taken by the robot will vary slightly due to variations in velocity, joint geometries, and point spatial locations. This difference in paths is difficult to predict and therefore can create a potential safety hazard to personnel and equipment.

Controlled Path

The path or mode of movement ensures that the end of the robot's arm will follow a predictable (controlled) path and orientation as the robot travels from point to point. The numerical coordinate transformations required for this hardware management are calculated by the robot's control system computer. Observations that result from this type of programming are less likely to present a hazard to personnel and equipment.

Continuous Path

A robot whose path is controlled by storing a large number or a close succession of spatial points, in memory during a teaching sequence is a continuous path controlled robot. During this time and while the robot is being moved, the coordinate points in space of each axis are continually monitored on a fixed time base (e.g., 60 or more times per second) and placed into the control system's computer memory. When the robot is placed in the automatic mode of operation, the program is replayed from memory and a duplicate path is generated.

Robot Components

Industrial robots have four major components: the mechanical unit, power source, control system, and tooling. Figure 14-2 shows a robotic system.

1. Mechanical unit—

    a. The robot's manipulative arm is the mechanical unit. This mechanical unit is also comprised of a fabricated structural frame with provisions for supporting mechanical linkage and joints, guides, actuators (linear or rotary), control valves, and sensors. The physical dimensions, design, and weight-carrying ability depend on application requirements.

2. Power sources—

    a. Energy is provided to various robot actuators and their controllers as pneumatic, hydraulic, or electrical power. The robot's drives are usually mechanical combinations powered by these types of energy, and the selection is usually based upon application requirements. For example, pneumatic power (low-pressure air) is used generally for low weight carrying robots.

    b. Hydraulic power transmission (high-pressure oil) is usually used for medium to high force or weight applications, or where smoother motion control can be achieved than with pneumatics. Consideration should be given to potential hazards, such as fires from leaks when petroleum-based oils are used.

    c. Electrically powered robots are the most prevalent in industry. Either AC or DC electrical power is used to supply energy to electromechanical motor-driven actuating mechanisms and their respective control systems. Motion control is much better and, in an emergency, an electrically powered robot can be stopped or powered down more safely and faster than those with either pneumatic or hydraulic power.

3. Control systems—

    a. Most industrial robots incorporate computer or microprocessor-based controllers. These perform computational functions and interface with control sensors, grippers, tooling, and other peripheral equipment. The control system also performs sequencing and memory functions associated with communication and interfacing for on-line sensing, branching, and integration of other equipment.

    b. Controller programming may be done on-line or from remote, off-line control stations. Programs may be on cassettes, floppy disks, internal drives, or in memory; and may be loaded or downloaded by cassettes, disks, or telephone modem. Robot controllers can have self-diagnostic capability, which can reduce the downtime of robot systems. Some robot controllers have sufficient computational ability, memory capacity, and input/output capability to serve as system controllers for other equipment and processes. In addition, the robot controller may be in a control hierarchy in which it receives instructions and reports positions or gives directions. Robot manufacturers typically use proprietary languages for programming robot controllers and systems.

4. Tooling—

    a. Tooling is manipulated by the robot to perform the functions required for the application. Depending on the application, the robot may have one functional capability, such as making spot welds or spray painting. These capabilities may be integrated with the robot's mechanical system or may be attached at the robot's wristed effector interface. Alternatively, the robot may use multiple tools that may be changed manually (as part of setup for a new program) or automatically during a work cycle.

b.   Tooling and objects that may be carried by a robot's gripper can significantly increase the envelope in which objects or humans may be struck. Tooling manipulated by the industrial robot and carried objects can cause more significant hazards than the motion of bare robotic system. Hazards added by tooling should be addressed as part of the risk assessment.

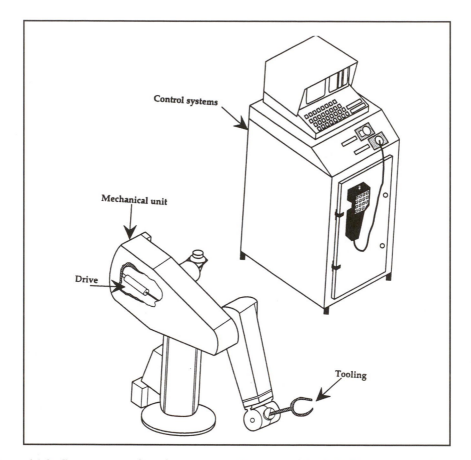

Figure 14-2. Components of a robotic system. Courtesy of the U.S. Department of Energy

Installation of Robots

Manufactured, Remanufactured, and Rebuilt Robots

All robots should meet minimum design requirements to ensure safe operation by the user. Consideration needs to be given to a number of factors in designing and building robots to industry standards. If older or obsolete robots are rebuilt or remanufactured, they should be upgraded to conform to current industry standards.

Every robot should be designed, manufactured, remanufactured, or rebuilt with safe design and manufacturing considerations. Improper design and manufacture can result in hazards to personnel if minimum industry standards are not conformed to on mechanical components, controls, methods of operation, and other required information necessary to insure safe and proper operating procedures. To ensure that robots are designed, manufactured, remanufactured, and rebuilt to ensure safe operation, it is recommended that they comply with section four of the ANSI/RIA R15.06-1992 standard for the Manufacturing, Remanufacture, and Rebuilding of Robots.

Installation

A robot or robotic system should be installed by the users in accordance with the manufacturer's recommendations and in compliance with acceptable industry standards. Temporary safeguarding devices and practices should be used to minimize the hazards associated with the installation of new equipment. The facilities, peripheral equipment, and operating conditions which should be considered are:

- Installation specifications.
- Physical facilities.
- Electrical facilities.
- Action of peripheral equipment integrated with the robot.
- Identification requirements.
- Control and emergency stop requirements.
- Special robot operating procedures or conditions.

To ensure safe operating practices and safe installation of robots and robot systems, it is recommended that the minimum requirements of section five of the ANSI/RIA R15.06-1992, Installation of Robots and Robot Systems be followed. In addition, OSHA's Lockout/Tagout standards (29 CFR 1910.147 and 1910.333) must be followed for servicing and maintenance.

Robot Programming by Teach Methods

Robots perform tasks for a given application by following a programmed sequence of directions from the control system. The robot's program establishes a physical relationship between the robot and other equipment. The program consists of a sequence of positions for the axes of movement and any end-elector operation, path information, timing, velocities, sensor data reading, external data-source reading, and commands or output to externally connected systems. The program may be taught by manually commanding the robot to learn a series of positions and operations (such as gripper closing) that collectively compose the work cycle. The robot converts these positions and operations into its programming language.

Alternatively, robot programming can be inputted directly in its programming language at a terminal, which may be the robot's controller or a separate computer. Robot programming generally needs verification and some modifications. This procedure is called program touchup. It is normally done in the each mode of operation with the teacher manually leading the robot through the preprogrammed steps. Three different teaching or programming techniques are lead-through, walk-through, and off-line programming. A description of each is provided below.

Lead-Through Programming/Teaching

Lead-through programming usually uses a teach pendant. This allows the teacher to direct the robot through a series of positions and to enter associate commands and other information, such as velocities. A human teaches the positions. The robot's controller generates the programming commands to move the robot between positions when the program is played. When using this programming technique, the teacher may need to enter the robot's working envelope. This introduces a high potential for accidents because safeguarding devices may have to be deactivated to permit such entry.

Only the teach pendant may be used to program a robot, or it may be used with an additional programming console and/or the robot's controller.

Walk-Through Programming/Teaching

The teacher physically moves ("walks") the robot through the desired positions within the robot's working envelope. During this time, the robot's controller may scan and store coordinate values on a fixed-time interval basis. These values and other functional information are replayed in the automatic mode. This may be at a different speed than that used in the walk-through.

This type of walk-through programming uses triggers on manual handles that move the robot. When the trigger is depressed, the controller remembers the position. The movement between these points, when the program is played, would then be generated by the controller.

Walk-through methods of programming require the teacher to be within the robot's working envelope with the robot's controller energized at least in the position sensors. This may also require that safeguarding devices be deactivated.

Off-line Programming

Off-line programming uses a remote programming computer. The programmer establishes the required sequence of functional and positional steps. The program is transferred to the robot's controller by disk, cassette, or network link. Typically, positional references are established on the robot to calibrate or transform the coordinates used in the remote programming for the actual setup.

Degrees of Freedom

"Degrees of freedom" refers to the directions of motion inherent in the design of robot mechanical systems. Each actuator usually causes rotary or linear motion along an axis. The number of axes is normally the same as the number of degrees of freedom for the robot. Motions of actuators in end effectors, such as closing grippers or the motion of a drill, do not constitute additional degrees of freedom. Depending on the robot's geometry, motion of one or more axes may be required for the robot to move to a new location in space. (See Figure 14-3.)

There are usually three axes, which move the robot's wrist-end effector interface to a position in space. Additional axes provide for rotation at that point to permit flexibility in orientation. One, two, or three rotating axes at the wrist may be used, depending on the requirements of the application and the robotic system's sophistication. Six degrees of freedom are commonly available with articulated arm and gantry robots. Four degrees of freedom are typical with the selective compliance assembly robot arm (SCARA) configuration. Seven or more axes are used for some special applications. An example is an articulated arm-welding robot required to work on the far side of an automobile body. A seventh degree of freedom is also added by putting a six-degree-of-freedom robot on ways or rails, allowing for an extended longitudinal range.

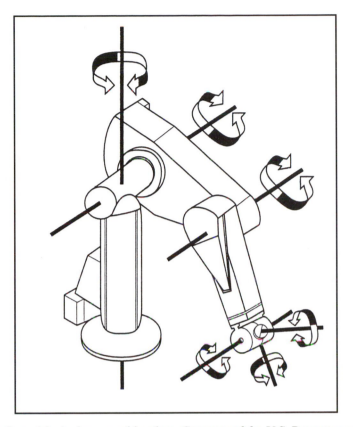

Figure 14-3. Robot with six degrees of freedom. Courtesy of the U.S. Department of Energy

PROTECTIVE DEVICES

The proper selection of an effective robotics safeguarding system should be based upon a hazard analysis of the robotics system application and required operations. Prior to safeguarding a robot or robot system, it is recommended that a risk assessment be performed and appropriate safeguards are employed to reduce or eliminate the risks of injury from potential hazards. Among the factors to be considered are tasks a robot is programmed to perform, start-up and programming procedures, environmental conditions, locations and installation requirements, possible human errors, scheduled and unscheduled maintenance, possible robot and system malfunctions, normal mode of operation functions, and all personnel functions and duties.

Safeguarding of personnel with varying job functions requires special considerations as far as the types of safety measures that can be employed. Operators who oversee the robot and its system in the automatic mode are usually exposed to fewer hazards because all safeguarding devices employed are activated, and their operators are usually outside of the robot's restricted envelope (space) and safeguarded area. Personnel who teach or perform programming and maintenance functions, or attend continuous operation functions work with robot system safeguards deactivated and with the robot in manual or teach mode.

An effective safeguarding system protects not only operators, but also engineers, programming and maintenance personnel, and any others who, by nature of their function, could be exposed to hazards associated with a robot's operation. A combination of safeguarding methods may be used. Redundancy and backup systems are especially recommended, particularly if a robot or a robot system is operating under hazardous conditions or handling hazardous materials. Safeguarding devices employed should not constitute or act as a hazard or obstruct the vision of operators.

Personnel should be safeguarded from hazards associated with the restricted envelope through the use of limiting devices and one or more of the safeguarding devices listed below:

- Mechanical limiting devices, such as emergency stop buttons and enabling switches.
- Nonmechanical limiting devices.
- Presence sensing safeguarding devices. (See Figure 14-4.)
- Fixed barriers (recommended at least two meters in height).
- Interlock barriers.

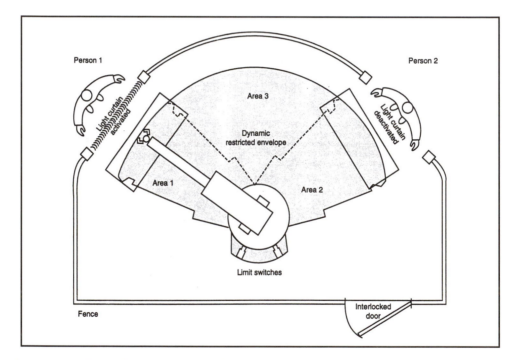

Figure 14-4. Example of a robotic work area with safety devices. Courtesy of the U.S. Department of Energy

Awareness devices are usually used with other safeguarding devices. Typical awareness devices are chalk or roped barriers with supporting stations or flashing lights, signs, whistles, and horns.

An amber light should be installed on the robot to be conspicuous from anywhere around the robot. This light must be on while the robot is energized, signifying it is "live," even during periods when the robot is not moving.

Safeguarding Methods

Proper selection of an effective robotic safeguarding system should be based upon a hazard analysis of the robot system's use, programming, and maintenance operations. Among the factors to be considered are the tasks a robot will be programmed to perform, start-up and command or programming procedures, environmental conditions, location and installation requirements, possible human errors, scheduled and unscheduled maintenance, possible robot and system malfunctions, normal mode of operation, and all personnel functions and duties.

An effective safeguarding system protects not only operators but also engineers, programmers, maintenance personnel, and any others who work on or with robot systems and could be exposed to hazards associated with a robot's operation. A combination of safeguarding methods may be used. Redundancy and backup systems are especially recommended, particularly if a robot or robot system is operating in hazardous conditions or handling hazardous materials. The safeguarding devices employed should not themselves constitute or act as a hazard or curtail necessary vision or viewing by attending human operators.

The operating envelope of the robot should be physically restricted. This is accomplished by using some form of mechanical stop(s) able to withstand the force of momentum of the robot traveling at maximum speed and carrying a full load. (See Figure 14-5.)

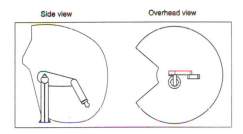

Illustration of maximum envelope

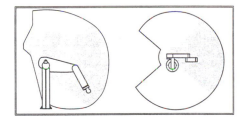

Illustration of restricted envelope

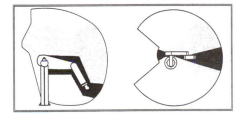

Illustration of operating envelope

Figure 14-5. Illustrates the robot operating envelope. Courtesy of the U.S. Department of Energy

Safeguarding the teacher or the robot programmer requires special consideration to minimize as many hazards as possible. During the teach mode of an operation, the teacher is responsible for controlling the robot and associated equipment and for being familiar with the operations to be programmed, system interfacing, and the control functions of the robot and other equipment. When systems are large and complex, it can be easy for improper functions to be activated or sequenced. Because the teacher can be within the robot's restricted envelope, mistakes in programming can result in unintended movement or actions. For this reason, a restricted speed of 150 millimeters per second (six inches per second) has been placed on any part of the robot. This slower speed should minimize potential injuries to the teacher if an inadvertent action or movement occurred. Several other safeguards are suggested to reduce the hazards associated with how training personnel use the robots. Refer to section 6.5 of the ANSI/RIA R15.06-1986 standard.

To prevent hazards to the system, the robot systems operator should ensure that all appropriate safeguards are established for all robot operations. Because the robot will be in the automatic mode of operation, all safeguarding devices must be activated. At no time should any part of the operator's body be within the robot's safeguarded area. For additional information on operator safeguarding, see section 6.6 of the ANSI/RIA R15.06-1986 standard.

The attended continuous operation mode permits a person to be in or near the robot's restricted envelope to evaluate or check the robot's motion or other operations being performed. The robot should be moving at a slow speed, and the operator should have the robot in the teach mode and be in full control of all operations being performed. Other safeguarding requirements are suggested. (See section 6.7 of the ANSI/RIA R15.06-1986 standard.)

The safeguarding of the job functions of maintenance and repair personnel is difficult for many reasons. These personnel are exposed to unknown hazards because part of their responsibilities includes troubleshooting faults and problems with the robot, controller, tooling, and other associated equipment.

During maintenance and repair, the robot should be placed in the manual or teach mode, to protect maintenance personnel who perform their work within the safeguarded area and the robot's restricted envelope.

Review of Safeguards

For the planning stage, installation, and subsequent operation of a robot or robot system, one should consider the following safeguards:

1. Risk assessment—At each stage of development of the robot and robot system a risk assessment should be performed. There should be different system and personnel safeguarding requirements at each stage. The appropriate level of safeguarding determined by the risk assessment should be applied. In addition, risk assessments for each stage of development should be documented for future reference.

2. Safeguarding devices—Personnel should be safeguarded from hazards associated with the restricted envelope (space) through the use of one or more safeguarding devices:

    a. Mechanical limiting devices.

    b. Nonmechanical limiting devices.

    c. Presence-sensing safeguarding devices.

    d. Fixed barriers (which prevent contact with moving parts).

    e. Interlocked barrier guards.

5. Awareness devices—Typical awareness devices include chain or rope barriers with supporting stanchions or flashing lights, signs, whistles, and horns. They are usually used in conjunction with other safeguarding devices.

6. Safeguarding the teacher—Special consideration must be given to the teacher or person who is programming the robot. During the teach mode of operation, the person performing the teaching has control of the robot and associated equipment. The teacher should be familiar with the operations to be programmed, system interfacing, and control functions of the robot and other equipment. When systems are large and complex, it can be easy to activate improper functions or sequence functions improperly. Since the person doing the training can be within the robot's restricted envelope, such mistakes can result in accidents. Mistakes in programming can result in unintended movement or actions with similar results. For this reason, a restricted speed of 250 millimeters per second or ten inches per second should be placed on any part of the robot during training to minimize potential injuries to teaching personnel.

7. Operator safeguards —The system operator should be protected from all hazards during operations performed by the robot. When the robot is operating automatically, all safeguarding devices should be activated, and at no time should any part of the operator's body be within the robot's safeguarded area.

8. Attended continuous operation—When a person is permitted to be in or near the robot's restricted envelope to evaluate or check the robot's motion or other operations, all continuous operation safeguards must be in force. During this operation, the robot should be at slow speed, and the operator should have the robot in the teach mode and be fully in control of all operations.

9. Maintenance and repair personnel—Safeguarding maintenance and repair personnel is very difficult because their job functions are so varied. Troubleshooting faults or problems with the robot, controller, tooling, or other associated equipment is just part of their job. Program touchup is another of their many jobs, as is scheduled maintenance, and adjustments of tooling, gauges, and recalibration.

   While maintenance and repair are being performed, the robot should be placed in the manual or teach mode, and the maintenance personnel should perform their work within the safeguarded area and within the robot's restricted envelope. Additional hazards are present during this mode of operation because the robot system safeguards are not operative.

10. Maintenance—Maintenance should occur during the regular and periodic inspection program for a robot or robot system. An inspection program should include, but not be limited to, the recommendations of the robot manufacturer and manufacturer of other associated robot system equipment such as conveyor mechanisms, parts feeders, tooling, gages, sensors, and the like.

    These recommended inspection and maintenance programs are essential for minimizing hazards from component malfunction, breakage, and unpredicted movements or actions by the robot or other systems equipment. To ensure proper maintenance, it is recommended that periodic maintenance and inspections be documented along with the identity of personnel performing these tasks.

11. Safety training—Personnel who program, operate, maintain, or repair robots or robot systems should receive adequate safety training, and they should be

able to demonstrate their competence to perform their jobs safely. Employers can refer to OSHA's publication 2254 (Revised), "Training Requirements in OSHA Standards and Training Guidelines."

12.   General requirements—To ensure minimum safe operating practices and safeguards for robots and robot systems covered in sections of ANSI/RIA R15.06-1992. The sections which must be considered are:

a.   Section 6—Safeguarding Personnel.

b.   Section 7—Maintenance of Robots and Robot Systems.

c.   Section 8—Testing and Start-up of Robots and Robot Systems.

d.   Section 9—Safety Training of Personnel.

Maintenance Procedures

The user of a robot or robot system should establish a regular and periodic inspection and maintenance program to ensure safe equipment operations. This program should include, but not be limited to, the recommendations of the robot manufacturer and the manufacturers of other associated robot system equipment such as conveyer mechanisms, part feeders, tooling, gauges, and sensors. These recommended maintenance programs are essential for minimizing the hazards that can result from component malfunction, breakage, and unpredicted movements or actions of the robot or other system equipment.

To ensure that robots are safely and adequately maintained, it is recommended that periodic maintenance be conducted and documented, including the names of the personnel who perform maintenance and the names of the independent verifiers. It is important that maintenance not be performed on energized robots and a lockout/tagout procedure should be employed to prevent errant release of energy.

Operator Knowledge

Personnel involved with the installation and operation of robots must have knowledge of, and access to, information on the robotic system as necessary. The following documentation should be provided by the robot manufacturer or created by the user:

• Function and location of all controls.

• Lifting procedures and precautions.

• Manufacturers' system-specific safety related information.

• Operating instructions.

• Maintenance and repair procedures, including lockout and tagout procedures.

• Robot system testing and start-up procedures, including initial start-up procedures.

• Installation instructions.

• Information on special environmental requirements.

• Electrical requirements.

• System-specific safety documentation, including risk assessment documentation.

• System-specific robot safety training lesson plans and associated materials.

• System-specific maintenance, failure, mishap, and training records, including calibration checking and calibration procedures.

TRAINING

All personnel involved with the installation, maintenance, and operation of robot systems should be trained regarding the hazards, safe operating procedures, and safeguards for robotic work areas.

Technical Training

Robotic system safety depends on passive and active barriers and restraints, built-in enabling and disenabling subsystems, procedures, and adherence to procedures by personnel for safe and effective use of the systems. Management should ensure that personnel who program, operate, maintain, or repair robots or robot systems are adequately trained to perform their assigned functions. This training must include safety aspects of the subject systems and their individually assigned functions. Technical training may be conducted at robotic system suppliers' facilities or at the user's facility. Technical training may be provided on the job by formally trained personnel and documented in the employee's training record. Supervised on-the-job training is recommended after all technical training. Performance based on training is recommended.

Initial Safety Training

Safety training is required in addition to technical training. Safety training is tailored to the actual installation and facility and to the functions that are performed at that facility. This same training should be provided for all personnel responsible for robotic systems. Management should maintain a list of individuals who have successfully completed safety training and the dates thereof. The robotic facility must be made available for safety training, which must include the following.

- Understanding of the robot system's maximum, restricted, and working envelopes, and zones or areas where access is limited and/or controlled.

- A review of hazards and their potential consequences.

- Review of safeguards: passive barriers and features limiting robot system movement, passive and active warning systems, and integral features of the system.

- Use and location of emergency stops.

- Safe use of any teach pendant and other controls that operate within the maximum envelope.

- Review of safety-related procedures and manufacturers' safety recommendations.

- A description of the system's operational functions and the need for hazard and system reassessment if changes are made to these functions.

- Recognition of off-normal and emergency situations and appropriate responses.

- Individual responsibilities by name and position (including management, operational, technical, maintenance, and repair personnel), emergency contacts, and location where these are posted.

- Identification of documentation associated with safety, and locations of reference copies available to personnel at the facility.

Refresher Training

Refresher safety training shall be provided to personnel who program, operate, maintain, or repair robots or robot systems. At minimum, this training should be provided annually, when operational or safeguarding systems change, or when management determines that additional training is required. All training materials must be updated to reflect the robotic system in its current configuration. Refresher training should cover the same material as the initial safety training.

APPLICABLE REGULATIONS AND STANDARDS

Although OSHA has not developed a specific regulation regarding industrial robots, there are OSHA regulations which have application to robot safety. Robots or robotic systems must comply with the following regulations: Occupational Safety and Health Administration, OSHA 29 CFR 1910.333, Selection and Use of Work Practices, and OSHA 29 CFR Part 1910.147, The Control of Hazardous Energy (Lockout/Tagout). Many other agencies have developed standards which are applicable to industrial robots. (See Table 14-1.)

Table 14-1

Standards and Codes for Industrial Robots

	Group	Standard	Subject
1.	ANSI/RIA	R1506-1986	American national standards for industrial robots and robot systems.
2.	BSR/RIA	BSR/RIA R15-06-19XX	Proposed standard for industrial robots and robot systems.
3.	ANSI/RIA	R15.02-1990	American national standard human engineering design criteria for hand-held robot control pendants.
4.	OSHA	Pub. 2254 (revised)	Training requirements in standards and training guidelines.
5.	NIOSH	Pub. 88-108	Safe maintenance guidelines for robotics workstations.
6.	OSHA	Pub. 8-1.3, 1987	Guidelines for robotics safety.
7.	OSHA	29 CFR 1910.147	Control of hazardous energy source (lockout/tagout final rule).
8.	OSHA	29 CFR 1910 - Subpart O	Machinery and machine guarding.
9.	AFOSH	127-12, 1991	Occupational safety machinery.

ANSI/RIA =   American National Standards Institutes/Robotics Industrial Association.
BSR/RIA =   Bureau of Standards Review/Robotics Industrial Association.
NIOSH   =   National Institute for Occupational Safety and Health.
OSHA   =   Occupational Safety and Health Administration.
AFOSH   =   Department of the Air Force.

OTHER ROBOTIC SYSTEMS NOT COVERED

Not all robot applications have been covered in this chapter, even though most of the safety principles would be applicable to any other robot systems like the following:

- Service robots are machines that extend human capabilities.

- Automatic guided-vehicle systems are advanced material-handling or conveying systems involving a driverless vehicle that follows a guide-path.

- Undersea and space robots include, in addition to the manipulator or tool that actually accomplishes a task, the vehicles or platforms that transport the tools to the site. These vehicles are called remotely operated vehicles (ROVs) or autonomous undersea vehicles (AUVs); the feature that distinguishes them is, respectively, the presence or absence of an electronics tether that connects the vehicle and surface control station.

- Automatic storage and retrieval systems are storage racks linked through automatically controlled conveyors and an automatic storage and retrieval machine or machines that ride on floor-mounted guide rails and power-driven wheels.

- Automatic conveyor and shuttle systems are comprised of various types of conveying systems linked together with various shuttle mechanisms for the prime purpose of conveying materials or parts to prepositioned and predetermined locations automatically.

- Teleoperators are robotic devices comprised of sensors and actuators for mobility and/or manipulation and are controlled remotely by a human operator.

- Mobile robots are freely moving automatic programmable industrial robots.

- Prosthetic robots are programmable manipulators or devices for missing human limbs.

- Numerically controlled machine tools are operated by a series of coded instructions comprised of numbers, letters of the alphabet, and other symbols. These are translated into pulses of electrical current or other output signals that activate motors and other devices to run the machine.

REFERENCES

American National Standards Institute, Inc. *Industrial Robots and Robot Systems - Safety Requirements (ANSI/ RIA R15.06-1992)*. New York, NY, 1992.

Department of Health and Human Services. National Institute for Occupational Safety and Health. *Request for Assistance in Preventing the Injury of Workers by Robots*. No. 85103. Morgantown, WV, 1985.

Department of Health and Human Services. National Institute for Occupational Safety and Health. *Safe Maintenance Guidelines for Robotic Workstations*. Technical Report Publication No. 880108. Morgantown, WV, 1988.

Underwriters Laboratories, Inc. *Standard for Industrial Robots and Robotic Equipment*. ANSI/UL1740, Northbrook, IL, 1996.

United States Department of Energy. *OSH Technical Reference Manual*. Washington, DC, 1993.

United States Department of Labor. Occupational Health and Safety Administration. *Guideline for Robotics Safety: OSHA Instruction Publication No. 8-1.3*. Washington, DC, 1987.

United States Department of Labor. Occupational Safety and Health Administration. *OSH Technical Manual: Section IV-Chapter 4*. Industrial Robots and Robot System Safety. Washington, DC, 1992.

CHAPTER 15

STORAGE AND HOUSEKEEPING

Bad example of storage and housekeeping

Material storage and housekeeping are components which contribute to the safety and health aspects of material handling. Most certainly proper and secure storage of materials factors into the safe handling of materials. Proper storage by the word itself suggests that housekeeping is an intended purpose of storage.

Housekeeping by general definition is a place for everything and everything in its place. Most safety professionals will support the concept that good housekeeping alone goes a long way toward accident prevention and not only from an appearance perspective is housekeeping important. Both proper storage and housekeeping facilitate the efficiency and productivity, which will be enhanced by the neat as well as proper storage of materials by expediting effective handling of materials of all kinds.

Storage is truly not a simple concept. It takes special care and specific techniques such as racks, pallets, bins, handling equipment, and procedures. The major reason that complexity of storage depends on the variable shapes of materials to be stored, which may range from crates to plastic pipe. (See Table 15-1.)

Table 15-1

Types of Materials to be Stored

bags	pipe
barrels	plate glass
blocks (concrete)	rebar
boxes	reels
cartons	rolls
drums	round objects
flat objects	scrap metal
heavy objects	sheet metal
irregular shapes	tanks
kegs	lumber

The physical parameters are not the only one aspect of materials, which must be considered when storing them. Many materials have potential hazards such as toxicity, corrosive, reactivity, explosive, combustible, and flammability, which may create the need for specially designed containers and storage procedures to assure the safety and health of those handling hazardous materials in the workplace. Storage and handling of loose or bulk materials must also be addressed since hazards may come from bins, hoppers, stockpiles, silos, and tanks.

This chapter emphasizes storage and housekeeping issues related to material handling and storage, flammable and combustible liquids, and compressed gases.

MATERIALS HANDLING AND STORAGE (SUBPART N OF 29 CFR 1910)

More employees are injured in industry while moving materials than while performing any other single function. In everyday operations, workers handle, transport, and store materials. They may do so by hand, by manually operated materials handling equipment, or by power operated equipment. The general guidelines for handling materials is found in the OSHA regulations 29 CFR 1910.176. A summary of those regulatory requirements follows. A material handling checklist is found in Figure 15-1, which provides guidance for compliance with existing regulations.

Use of Mechanical Equipment

Where mechanical handling equipment is used, sufficient safe clearance shall be allowed for aisles, at loading docks, through doorways, and whenever turns or passage must be made. Permanent aisles and passageways shall be appropriately marked.

Secure Storage

Storage of material shall not create a hazard. All stored materials stacked in tiers shall be stacked, blocked, interlocked, and limited in height so that they are secure against sliding or collapse.

Housekeeping

Storage areas shall be kept free from accumulation of materials that constitute hazards from tripping, fire, explosion or pest harborage. Vegetation control will be exercised when necessary.

MATERIALS HANDLING CHECKLIST

_____ Is material safely stored from slipping or falling?

_____ Do load bearing floors (except ground surface) with known and posted load limits exist?

_____ Are stored materials free from open sides and hoistways?

_____ Are bagged materials, stacked blocks, bricks and lumber appropriately stored by height and secured including tappering?

_____ Are pipe, structural steel, poles safely stored and blocked?

_____ Is used lumber stacked and stable without protruding nails?

_____ Is handling equipment inspected and removed from work area when not in use?

_____ Is all rigging equipment used properly and within load limits?

_____ Are chains inspected, in good repair, weight tagged and used safely?

_____ Is wire rope (cable) inspected, in good condition, knot and kink free, and used within safe limits?

_____ Are eyes made with U-bolts properly attached?

_____ Are fiber rope slings and lines in good condition, used safely?

_____ Are web slings in good condition, used within load limits?

_____ Are appropriate shackles used within load limit?

_____ Is waste and scrap dropped through chutes or openings that are clearly barricaded or posted?

_____ Is a scrap removal program in progress, and scrap not accumulating?

_____ Are oily rags, combustible materials, and flammable waste contained in closed fire retardant containers?

Figure 15-1. Material handling and storage checklist

STORAGE

Storage of materials can come in many varied forms. Some materials are stored in a warehouse using pallets or shelves as seen in Figure 15-2 and 15-3. Other storage areas have been constructed for bulk storage. These areas are often hoppers, silos, and storage structures. (See Figure 15-4.) Each storage enclosure has inherent safety hazards to be considered.

Figure 15-2. Using shelves for storage

Figure 15-3. Using pallets for storage

Figure 15-4. Storage structure to protect bulk road sand and salt

Silos

Silos of different types pose an engulfment hazard as well as an explosion hazard. Materials in the form of grain, solid chemicals, and other bulk materials are conveyed into the tops of silos. At times, when finely dispersed dust unites with an ignition source, the result can be a very explosive situation. Much material on explosive silos and elevators exists and is beyond the scope of this book. But all too often, workers are asked to enter silos to loosen materials caked on the walls or break bridged material loose. These environment may be oxygen deficient and should be tested prior to entry, but entrapment is the biggest hazard. Thus workers should be required to wear a safety harness and lifeline to prevent sliding into the bulk material and being buried. (See Figure 15-5.)

Hoppers

Hoppers are bulk storage units with slanted walls, constricted at the base. The main hazards are similar to those of silos, but the constricted base and small opening make entry through the top the only option to dislodge materials. This also poses an engulfment hazard and the steep walls make slipping almost an assured event. Thus, the use of a safety harness and lifeline is a must. (See Figure 15-6.)

Figure 15-5. Grain silos

Figure 15-6. Hopper used for storage and dispensing of loose materials

Stockpiles

Stockpiles are usually areas of high activity with the movement of many workers and varied types of equipment, as well as ever-present truck traffic. These areas are also hazardous since they are composed of unconsolidated materials, which have the potential to slip, fracture (crack), slide, slump, or bridge. This definitely increases the risk for engulfment of equipment and people. Materials in stockpiles tend to seek a natural angle of repose for the particular type of material; for example, sand's angle of repose is 34 degrees. Once material is removed from the stockpile, it will naturally move. This is why stockpiles should be trimmed or made safe. (See Figure 15-7.) Some of the rules to adhere to around stockpiles are as follows:

1. Operators of all equipment should be well-trained.

2. Trucks should dump away from the edge of a stockpile.

3. Dozers should never push a load over the edge, but should use the next load to bump the first load over the edge.

4. All equipment should work perpendicular to the edge of the stockpile, not parallel.

5. Berms shall be provided on the outer edges of elevated roadways and stockpiles.

6. Berms should not be depended upon to stop vehicles, trucks and mobile equipment.

7. Hand loading should be prohibited around stockpiles.

8. Workers should never position themselves between equipment and the toe (bottom) of the stockpile.

Figure 15-7. An example of a large stockpile

MATERIAL HANDLING AND STORAGE IN CONSTRUCTION (29 CFR 1926.250)

Materials handling accounts for approximately 40 percent of the incidents resulting in lost time that occur in the construction industry. These injuries are often a result of inadequate planning, administration of safety policies, and/or engineering approaches. Therefore, in an effort to reduce workplace injuries, the following safe work procedures must be implemented and enforced at all construction projects.

Materials that are stored in tiers must be stacked, racked, blocked, interlocked, and otherwise secured to prevent sliding, falling, or collapsing. Noncompatible materials are to be segregated in storage. Bagged materials must be stacked by stepping back the layers and cross-keying the bags at least every ten bags high. Brick stacks are not to be more than seven feet in height. When a loose brick stack reaches a height of four feet, it is to be tapered back two inches in every foot of height above the four-foot level. When masonry blocks are stacked higher than six feet, the stack is to be tapered back one-half block per tier above the six-foot level.

Lumber is to be stacked on a level surface and solidly supported sills, and where it can be stable and self-supporting. Lumber piles must not exceed 20 feet in height, except lumber that is to be handled manually, which may not to be stacked more than 16 feet high. Any used lumber must have all nails withdrawn before stacking. Structural steel, poles, pipe, bar stock, and other cylindrical materials, unless racked, are to be stacked and blocked so as to prevent spreading or tilting. (See Figure 15-8.)

Materials stored inside buildings that are under construction are not to be placed within six feet of any hoistway or inside floor openings, nor within ten feet of an exterior wall which does not extend above the top of the material stored. The maximum safe load limits of floors within buildings and structures, in pounds per square foot, must be conspicuously posted in all storage areas, except for floors or slabs on a grade. The maximum safe loads are not to be exceeded. Materials are not to be stored on scaffolds or runways in excess of supplies needed for immediate operations.

Figure 15-8. Securely stacked cylindrical material

Also, all aisles and passageways should be kept clear to provide for the free and safe movement of material handling equipment or employees, and these are to be kept in good repair. When a difference in road or working levels exists, means such as ramps, blocking, or grading are to be used to ensure the safe movement of vehicles between the two levels.

Any time a worker is required to work on stored material in silos, hoppers, tanks, and similar storage areas, they are to be equipped with personal fall arrest equipment which meets the requirements of Subpart M.

"Housekeeping" storage areas are to be kept free from accumulating materials that constitute hazards from tripping, fire, explosion, or pest harborage. Vegetation control is to be exercised when necessary.

When portable and powered dockboards are used, they should be strong enough to carry the load imposed on them. Portable dockboards are to be secured in position, either by being anchored or equipped with devices, which will prevent their slipping. Handholds, or other effective means, are to be provided on portable dockboards to permit safe handling. Positive protection must be provided to prevent railroad cars from being moved while dockboards or bridge plates are in position.

When handling materials, do not attempt to lift or move a load that is too heavy for one person. Get help. Attach handles or holders to the load to reduce the possibility of pinching or smashing fingers. Wear protective gloves and clothing (i.e., aprons), if necessary, when handling loads with sharp or rough edges. When pulling or prying objects, be sure you are properly positioned.

During the weekly "tool-box" meetings, employees should receive instructions on proper materials handling practices so that they are aware of the following types of injuries associated with manual handling of materials:

- Strains and sprains from lifting loads improperly, or from carrying loads that are too heavy or large.

- Fractures and bruises caused by dropping or flying materials, or getting hands caught in pinch points.

- Cuts and abrasions caused by falling materials which have been improperly stored, or by cutting securing devices incorrectly.

Engineering controls should be used, if feasible, to redesign the job so that the lifting task becomes less hazardous. This includes reducing the size or weight of the object lifted, changing the height of a pallet or shelf, or installing a mechanical lifting aid.

Fire Prevention — Storage (29 CFR 1926.151)

Internal combustion engine-powered equipment shall be so located that exhausts are well away from combustible materials. When exhausts are piped to the outside of a building under construction, a clearance of at least six inches is to be maintained between such piping and combustible material.

Smoking is prohibited at or in the vicinity of operations which constitute a fire hazard, and signs are to be conspicuously posted: "No Smoking" or "Open Flame."

Portable battery powered lighting equipment used in connection with the storage, handling, or use of flammable gases or liquids is to be of the type approved for the hazardous locations.

The nozzle of air, inert gas, and steam lines or hoses, when used in the cleaning or ventilation of tanks and vessels that contain hazardous concentrations of flammable gases or vapors, are to be bonded to the tank or vessel shell. Bonding devices are not to be attached or detached in hazardous concentrations of flammable gases or vapors.

No temporary building shall be erected where it will adversely affect any means of exit. Temporary buildings, when located within another building or structure, must be of either noncombustible construction, or combustible construction having a fire resistance of not less than one hour.

Temporary buildings, located other than inside of another building and not used for storage or handling, or the use of flammable or combustible liquids, flammable gases, explosives, blasting agents, or similar hazardous occupancies, are to be located at a distance of not less than ten feet from another building or structure. Groups of temporary buildings, not exceeding 2,000 square feet in aggregate, are, for the purpose of this writing, to be considered a single temporary building.

Combustible materials are to be piled with due regard to the stability of piles, and in no case higher than 20 feet. Driveways between and around combustible storage piles are to be at least 15 feet wide and maintained free from the accumulation of rubbish, equipment, or other articles and materials. Driveways must be so spaced that a maximum grid system unit of 50 feet by 150 feet is produced. Entire storage sites are to be kept free from accumulation of unnecessary combustible materials. Weeds and grass shall be kept down, and a regular procedure must be provided for the periodic cleanup of the entire area. When there is danger of an underground fire, that land is not to be used for combustible or flammable storage. The method used for piling is, wherever possible, to be solid and in orderly and regular piles. No combustible material is to be stored outdoors within ten feet of a building or structure.

Indoor storage must not obstruct or adversely affect the means of exit, and all materials shall be stored, handled, and piled with due regard to their fire characteristics. Noncompatible materials which may create a fire hazard are to be segregated by a barrier that has a fire resistance of at least one hour. Material shall be piled to minimize the spread of fire internally, and to permit convenient access for firefighting. Stable piling must be maintained at all times. Aisle space is to be maintained to safely accommodate the widest vehicle that may be used within the building for firefighting purposes. A clearance of at least 36 inches is to be maintained between the top level of the stored material and the sprinkler deflectors. Clearance must be maintained around lights and heating units to prevent ignition of combustible materials. A clearance of 24 inches shall be maintained around the path of travel of fire doors, unless a barricade is provided, in which case no clearance is needed. Material must not be stored within 36 inches of a fire door opening.

FLAMMABLE AND COMBUSTIBLE LIQUIDS

The specific OSHA regulation for flammable and combustible liquids is found in 29 CFR 1910.106. The primary basis of this standard is the National Fire Protection Association's publication NFPA 30, Flammable and Combustible Liquids Code. This standard applies to the handling, storage, and use of flammable and combustible liquids with a flash point below 200 degrees Fahrenheit. There are two primary hazards associated with flammable and combustible liquids: explosion and fire. In order to prevent these hazards, this standard addresses the primary concerns of design and construction, ventilation, ignition sources, and storage.

Combustible Liquids

Combustible liquid means any liquid having a flash point at or above 100 degrees Fahrenheit (37.8 degrees Celsius). Combustible liquids shall be divided into two classes as follows:

1. Class II liquids shall include those with flash points at or above 100 degrees Fahrenheit (37.8 degrees Celsius) and below 140 degrees Fahrenheit (60 degrees Celsius), except any mixture having components with flash points of 200 degrees Fahrenheit (93.3 degrees Celsius) or higher, the volume of which makes up 99 percent or more of the total volume of the mixture.

2. Class III liquids shall include those with flash points at or above 140 degrees Fahrenheit (60 degrees Celsius). Class III liquids are subdivided into two subclasses:

    a. Class IIIA liquids shall include those with flash points at or above 140 degrees Fahrenheit (60 degrees Celsius) and below 200 degrees Fahrenheit (93.3 degrees Celsius), except any mixture having components with flash points of 200 degrees Fahrenheit (93.3 degrees Celsius), or higher, the total volume of which makes up 99 percent or more of the total volume of the mixture.

    b. Class IIIB liquids shall include those with flash points at or above 200 degrees Fahrenheit (93.3 degrees Celsius). This section does not regulate Class IIIB liquids. Where the term "Class III liquids" is used in this section, it shall mean only Class IIIA liquids.

When a combustible liquid is heated to within 30 degrees Fahrenheit (16.7 degrees Celsius) of its flash point, it shall be handled in accordance with the requirements for the next lower class of liquids.

Flammable Liquids

Flammable liquid means any liquid having a flash point below 100 degrees Fahrenheit (37.8 degrees Celsius), the total of which makes up 99 percent or more of the total volume of the mixture. Flammable liquids shall be known as Class I liquids. Class I liquids are divided into three classes as follows:

1. Class IA shall include liquids having flash points below 73 degrees Fahrenheit (22.8 degrees Celsius) and having a boiling point below 100 degrees Fahrenheit (37.8 degrees Celsius).

2. Class IB shall include liquids having flash points below 73 degrees Fahrenheit (22.8 degrees Celsius) and having a boiling point at or above 100 degrees Fahrenheit (37.8 degrees Celsius).

3. Class IC shall include liquids having flash points at or above 73 degrees Fahrenheit (22.8 degrees Celsius) and below 100 degrees Fahrenheit (37.8 degrees Celsius).

It should be mentioned that flash point was selected as the basis for classification of flammable and combustible liquids because it is directly related to a liquid's ability to generate vapor, i.e., its volatility. Since it is the vapor of the liquid, not the liquid itself, that burns, vapor generation becomes the primary factor in determining the fire hazard. The expression "low flash - high hazard" applies. Liquids having flash points below ambient storage temperatures generally display a rapid rate of flame spread over the surface of the liquid, since it is not necessary for the heat of the fire to expend its energy in heating the liquid to generate more vapor. (See Figure 15-9.)

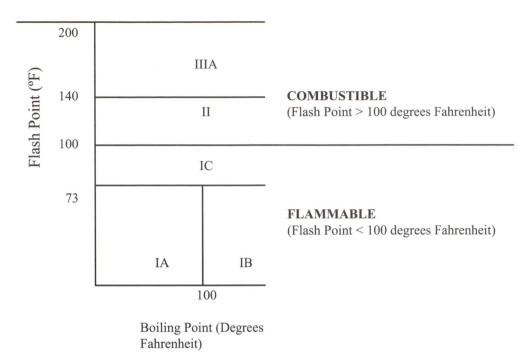

Figure 15-9. Classes of flammable and combustible liquids. Courtesy of the Occupational Safety and Health Administration

Container and Portable Tank Storage

This section applies only to the storage of flammable or combustible liquids in drums or other containers (including flammable aerosols) not exceeding 60 gallons individual capacity and portable tanks of less than 660 gallons individual capacity. A portable tank is a closed container that has a liquid capacity of over 60 gallons and is not intended for fixed installations. This section does not apply to the following:

- Storage of containers in bulk plants, service stations, refineries, chemical plants, and distilleries.
- Class I or Class II liquids in the fuel tanks of a motor vehicle, aircraft, boat, or portable or stationary engine.
- Flammable or combustible paints, oils, varnishes, and similar mixtures used for painting or maintenance when not kept for a period in excess of 30 days.
- Beverages when packed in individual containers not exceeding 1 gallon in size.

Design, Construction, and Capacity of Containers

Only approved containers and portable tanks may be used to store flammable and combustible liquids. Metal containers and portable tanks meeting the requirements of the Department of Transportation (DOT) (49 CFR 178) are deemed acceptable when containing products authorized by the DOT (49 CFR 173).

The latest version of NFPA 30, Flammable and Combustible Liquids Code, indicates that certain petroleum products may be safely stored within plastic containers if the terms and conditions of the following specifications are met:

- ANSI/ASTM D 3435-80—Plastic Containers (Jerry Cans) for Petroleum Products.

- ASTM F 852-86—Standard for Portable Gasoline Containers for Consumer Use.

- ASTM F 976-86—Standard for Portable Kerosene Containers for Consumer Use.

- ANSI/UL 1313-83—Nonmetallic Safety Cans for Petroleum Products.

This standard also requires portable tanks to have provision for emergency venting. Top-mounted emergency vents must be capable of limiting internal pressure under fire exposure conditions to ten pounds per square inch gauge or 30 percent of the bursting pressure of the tank, whichever is greater. Portable tanks are also required to have at least one pressure-activated vent with a minimum capacity of 6,000 cubic feet of free air at 14.7 pound per square inch absolute and 60 degrees Fahrenheit. These vents must be set to open at not less than five pounds per square inch gauge. If fusible vents are used, they shall be actuated by elements that operate at a temperature not exceeding 300 degrees Fahrenheit. Maximum allowable sizes of various types of containers and portable tanks are specified based on the class of flammable and combustible liquid they contain.

Design, Construction, and Capacity of Storage Cabinets

Not more than 60 gallons of Class I and/or Class II liquids, or not more than 120 gallons of Class III liquids, may be stored in an individual cabinet. This standard permits both metal and wooden storage cabinets. Storage cabinets shall be designed and constructed to limit the internal temperature to not more than 325 degrees Fahrenheit when subjected to a standardized ten-minute fire test. All joints and seams must remain tight and the door shall remain securely closed during the fire test. Storage cabinets must be conspicuously labeled, "Flammable—Keep Fire Away."

The bottom, top, door, and sides of metal cabinets shall be at least number 18 gauge sheet metal and double walled with 1 1/2-inches air space. The door must be provided with a three-point lock, and the doorsill shall be raised at least two inches above the bottom of the cabinet.

Design and Construction of Inside Storage Rooms

Construction is to comply with the test specifications included in NFPA 251-1969, Standard Methods of Fire Tests of Building Construction and Materials. Openings to other rooms or buildings must be provided with non-combustible liquid-tight raised sills or ramps at least four inches in height, or the floor in the storage area shall be at least four inches below the surrounding floor. Openings shall be provided with approved self-closing fire doors. The room shall be liquid-tight where the walls join the floor. A permissible alternate to the sill or ramp is an open-grated trench inside the room, which drains to a safe location. This method may be preferred if there is an extensive need to transfer flammable liquids into and out of the room by means of hand trucks.

Rating and Capacity

Storage in inside storage rooms shall comply with Table 15 -2.

Table 15 -2

Storage in Inside Rooms

Fire Protection Provided[1]	Fire Resistance	Maximum Floor Area (ft^2)	Total Allowable Quantities (gal/ft^2 floor area)
Yes	2 hr.	500	10
No	2 hr.	500	4*
Yes	1 hr.	150	5*
No	1 hr.	150	2

* These numbers are incorrectly shown in 29 CFR 1910.106.

[1] The fire protection system shall be a sprinkler, water spray, carbon dioxide, or other system.

Wiring

Electrical wiring and equipment located in inside storage rooms used for Class I liquids shall be approved under Subpart S, Electrical, for Class I, Division 2 Hazardous Locations; for Class II and Class III liquids, it shall be approved for general use.

Ventilation

Every inside storage room shall be provided with either a gravity or a mechanical exhaust ventilation system designed to provide for a complete change of air within the room at least six times per hour. Ventilation is vital to the prevention of flammable liquid fires and explosions. It is important to ensure that air flow through the system is constant and prevents the accumulation of any flammable vapors.

Storage

In every inside storage room, there shall be maintained an aisle at least three feet wide. Easy movement within the room is necessary in order to reduce the potential for spilling or damaging the containers and to provide both access for fire fighting and a ready escape path for occupants of the room, should a fire occur.

Containers over 30-gallons capacity shall not be stacked one upon the other. Such containers are built to DOT specifications and are not required to withstand a drop test greater than three feet when full. Dispensing shall be done only by an approved pump or self-closing faucet.

Storage Inside Buildings

<u>Egress</u>

Flammable or combustible liquids, including stock for sale, shall not be stored so as to limit use of exits, stairways, or areas normally used for the safe egress of people.

Office Occupancies

Storage must be prohibited except that which is required for maintenance and operation of equipment. Such material storage shall be kept in closed metal containers stored in a storage cabinet or in safety cans or in an inside storage room not having a door that opens into that portion of the building used by the public.

General Purpose Public Warehouses

Public warehouses have criteria for indoor storage of flammable and combustible liquids, which are confined in containers and portable tanks. Storage of incompatible materials that create a fire exposure (e.g., oxidizers, water-reactive chemicals, certain acids and other chemicals) is not permitted.

Warehouses or Storage Buildings

The last type of inside storage covered herein is storage in "warehouses or storage buildings." These structures are sometimes referred to as outside storage rooms. Practically any quantity of flammable and combustible liquid can be stored in these buildings provided that they are stored in a configuration consistent with the following:

1.   Containers in piles shall be separated by pallets or dunnage where necessary to provide stability and to prevent excessive stress on container walls.

2.   Stored material shall not be piled within three feet of beams or girders and shall be at least three feet below sprinkler deflectors or discharge orifices of water spray, or other fire protection equipment.

3.   Aisles of at least three feet in width shall be maintained to access doors, windows or standpipe connections.

Storage Outside Buildings

Requirements covering "storage outside buildings" are summarized in this paragraph. Associated requirements are given for storage adjacent to buildings. Also included are requirements involving controls for diversion of spills away from buildings and security measures for protection against trespassing and tampering. Certain housekeeping requirements are given, which relate to control of weeds, debris, and accumulation of unnecessary combustibles.

Fire Control

Suitable fire control devices, such as small hoses or portable fire extinguishers, shall be available at locations where flammable or combustible liquids are stored. At least one portable fire extinguisher having a rating of not less than 12-B units shall be located:

- Outside, but not more than ten feet from, the door opening into any room used for storage.

- Not less than ten feet, nor more than 25 feet, from any Class I or Class II liquid storage area located outside of a storage room but inside a building.

The reason for requiring that portable fire extinguishers be located a distance away from the storage room is that fires involving Class I and Class II flammable liquids are likely to escalate rapidly. If the fire is too close to the storage area, it may be impossible to get to it once the fire has started. Open flames and smoking shall not be permitted in flammable or combustible liquid storage areas.

Materials that react with water shall not be stored in the same room with flammable or combustible liquids. Many flammable and combustible liquid storage areas are protected by automatic sprinkler or water spray systems and hose lines. Consequently, any storage of water-reactive material in the storage area creates an unreasonable risk.

Industrial Plants

The OSHA regulations as applied to industrial plants are as follows:

- The use of flammable or combustible liquids is incidental to the principal business.

- Where flammable or combustible liquids are handled or used only in unit physical operations such as mixing, drying, evaporating, filtering, distillation, and similar operations which do not involve chemical reaction.

- This does not to chemical plants, refineries or distilleries.

Incidental Storage or Use of Flammable or Combustible Liquids

The application of 29 CFR 1910.106 is directed to those portions of an industrial plant where the use and handling of flammable or combustible liquids is only incidental to the principal business, such as paint thinner storage in an automobile assembly plant, solvents used in the construction of electronic equipment, and flammable finishing materials used in furniture manufacturing,

Containers

Flammable or combustible liquids shall be stored in tanks or closed containers. The quantity of liquid that may be located outside of an inside storage room or storage cabinet in a building or in any one fire area of a building shall not exceed:

- 25 gallons of Class IA liquids in containers.
- 120 gallons of Class IB, IC, II, or III liquids in containers.
- 660 gallons of Class 1B, 1C, II, or III liquids in a single portable tank.

Handling Liquids at Point of Final Use

Flammable liquids shall be kept in covered containers when not actually in use. Where flammable or combustible liquids are used or handled, except in closed containers, means shall be provided to dispose promptly and safely of leakage or spills. Flammable or combustible liquids shall be drawn from or transferred into vessels, containers, or portable tanks within a building only in the following manner:

- Through a closed piping system.
- From safety cans.
- By means of a device drawing through the top.
- From containers or portable tanks by gravity through an approved self-closing valve.

Transfer operations must be provided with adequate ventilation. Sources of ignition are not permitted in areas where flammable vapors may travel. Transferring liquids by means of air pressure on the container or portable tanks is prohibited. This may result in an overpressure, which could exceed what the container or tank could withstand. In addition, a flammable atmosphere could be created within the container or tank. This atmosphere would be particularly sensitive to ignition because of the increased pressure.

Flammable and Combustible Liquids (29 CFR 1926.152)

When storing or handling flammable and combustible liquids, only approved containers and portable tanks are to be used. (See Figure 15 -10.) Approved metal safety cans are to be used for the handling and use of flammable liquids in quantities greater than one gallon, except that this does not apply to those flammable liquid materials which are highly viscid (extremely hard to pour); they may be used and handled in original shipping containers. For quantities of one gallon or less, only the original container or approved metal safety cans can be employed for storage, use, and handling of flammable liquids.

Figure 15-10. Portable storage tank

Flammable or combustible liquids are not to be stored in areas used for exits, stairways, or areas normally used for the safe passage of people. No more than 25 gallons of flammable or combustible liquids are to be stored in a room outside of an approved storage cabinet. Quantities of flammable and combustible liquid in excess of 25 gallons shall be stored in an acceptable or approved cabinet, and must meet the following requirements:

Specially designed and constructed wooden storage cabinets, which meet unique criteria, can be used for flammable and combustible liquids. Approved metal storage cabinets are also acceptable. Cabinets are to be labeled in conspicuous lettering, "Flammable-Keep Fire Away." Not more than 60 gallons of flammable, or 120 gallons of combustible, liquids are to be stored in any one storage cabinet. Not more than three such cabinets may be located in a single storage area. Quantities in excess of this must be stored in an inside storage room. Inside storage rooms are to be constructed to meet the required fire-resistive rating for their use. Such construction must comply with the test specifications set forth in the Standard Methods of Fire Testing of Building Construction and Material (NFPA 251-1969).

Where an automatic extinguishing system is provided, the system must be designed and installed in an approved manner. Openings to other rooms or buildings must be provided with noncombustible liquid-tight raised sills or ramps at least four inches in height, or the floor in the storage area must be at least four inches below the surrounding floor. Openings are to be provided with approved self-closing fire doors. The room is to be liquid-tight where the walls join the floor. A permissible alternate to the sill or ramp is an open-grated trench inside the room, which drains to a safe location. Where other portions of the building, or other buildings are exposed, windows are to be protected as set forth in the Standard for Fire Doors and Windows, NFPA No. 80-1970, for Class E or F openings. Wood of at least one inch nominal thickness may be used for shelving, racks, dunnage, scuffboards, floor overlay, and similar installations. Materials which will react with water and create a fire hazard are not to be stored in the same room with flammable or combustible liquids.

Electrical wiring and equipment located in inside storage rooms are to be approved for Class I, Division One, Hazardous Locations. Every inside storage room is to be provided with either a gravity or a mechanical exhausting system. Such a system shall commence not more than 12 inches above the floor, and be designed to provide for a complete change of air within the room at least six times per hour. If a mechanical exhaust system is used, it must be controlled by a switch located outside of the door. The ventilating equipment and any lighting fixtures are to be operated by the same switch. An electric pilot light shall be installed adjacent to the switch, if flammable liquids are dispensed within the room. Where gravity ventilation is provided, the fresh air intake, as well as the exhaust outlet from the room, must be on the exterior of the building in which the room is located.

In every inside storage room there shall be maintained one clear aisle at least three feet wide. Containers over 30 gallons capacity are not to be stacked one upon the other. Flammable and combustible liquids in excess of that permitted in inside storage rooms are to be stored outside of buildings.

The quantity of flammable or combustible liquids kept in the vicinity of spraying operations is to be the minimum required for operations, and should ordinarily not exceed a supply for one day, or one shift. Bulk storage of portable containers of flammable or combustible liquids are to be in a separate, constructed building detached from other important buildings, or cut off in a standard manner.

Storage of containers (not more than 60 gallons each) shall not exceed 1,100 gallons in any one pile or area. Piles or groups of containers are to be separated by a five-foot clearance. Piles or groups of containers are not to be closer than 20 feet to a building. Within 200 feet of each pile of containers, there should be a 12-foot-wide access way to permit the approach of fire control apparatus. The storage area is to be graded in a manner so as to divert possible spills away from buildings or other exposures, or is to be surrounded by a curb or earth dike at least 12 inches high. When curbs or dikes are used, provisions must be made for draining off accumulations of ground or rain water, or spills of flammable or combustible liquids. Drains shall terminate at a safe location, and must be accessible to the operation under fire conditions.

Portable tanks are not to be closer than 20 feet from any building. Two or more portable tanks, grouped together, having a combined capacity in excess of 2,200 gallons, are to be separated by a five-foot-clear area. Individual portable tanks exceeding 1,100 gallons shall be separated by a five-foot-clear area. Within 200 feet of each portable tank, there must be a 12-foot-wide access way to permit the approach of fire control apparatus.

Portable tanks, not exceeding 660 gallons, are to be provided with emergency venting and other devices, as required by Chapters III and IV of NFPA 30-1969, The Flammable and Combustible Liquids Code.

Portable tanks in excess of 660 gallons must have emergency venting and other devices, as required by Chapters II and III of The Flammable and Combustible Liquids Code, NFPA 30-1969. At least one portable fire extinguisher, having a rating of not less than 20-B units, is to be located outside of, but not more than ten feet from, the door opening into any room used for storage of more than 60 gallons of flammable or combustible liquids, and not less than 25 feet, nor more than 75 feet, from any flammable liquid storage area located outside.

At least one portable fire extinguisher, having a rating of not less than 20-B:C units, is to be provided on all tanker trucks or other vehicles used for transporting and/or dispensing flammable or combustible liquids.

Areas in which flammable or combustible liquids are transferred at one time, in quantities greater than five gallons from one tank or container to another tank or container, are to be separated from other operations by 25-feet distance, or by construction having a fire resistance of at least one hour. Drainage or other means are to be provided to control spills. Adequate natural or mechanical ventilation must be provided to maintain the concentration of flammable vapor at or below ten percent of the lower flammable limit.

Transfer of flammable liquids from one container to another is done only when containers are electrically interconnected (bonded), when flammable or combustible liquids are drawn from, or transferred into vessels, containers, or tanks within a building or outside. Transferring should only be done through a closed piping system, from safety cans, by means of a device drawing through the top, or from a container, or portable tanks, by gravity or pump, through an approved self-closing valve. Transferring by means of air pressure on the container or portable tanks is prohibited. Any dispensing units are to be protected against collision damage. Dispensing devices and nozzles for flammable liquids are to be of an approved type. The dispensing hose must be an approved type; also the dispensing nozzle must be an approved automatic-closing type without a latch-open device. Clearly identified and easily accessible switches are to be provided at a location remote from dispensing devices, in order to shut off the power to all dispensing devices in the event of an emergency.

Heating equipment, of an approved type, may be installed in the lubrication or service area where there is no dispensing or transferring of flammable liquids, provided the bottom of the heating unit is at least 18 inches above the floor and is protected from physical damage. Heating equipment installed in lubrication or service areas where flammable liquids are dispensed is to be of an approved type for garages, and is to be installed at least eight feet above the floor. No smoking or open flames are permitted in the areas used for fueling, servicing fuel systems for internal combustion engines, and receiving or dispensing flammable and combustible liquids. Conspicuous and legible signs prohibiting smoking are to be posted.

The motors of all equipment being fueled are to be shut off during the fueling operation. Each service or fueling area is to be provided with at least one fire extinguisher having a rating of not less than 20-B:C, and located so that an extinguisher will be within 75 feet of each pump, dispenser, underground fill pipe opening, and lubrication or service area. More information regarding specifics on flammable and combustible liquid storage tanks can be found in 29 CFR 1926.152. Also, further information on Fixed Extinguishing Systems (1926.156), Fixed Extinguishing Systems, Gaseous Agent (1926.157), and Fire Detection System (1926.158) can be found in 29 CFR 1926.

Summary

The regulation for the general industry (29 CFR 1910.106) and construction (29 CFR 1926.152) are very lengthy and it seemed to serve no purpose for this particular book to develop an extensive regulatory checklist. Thus, a brief discussion of each of these regulations has been put forth along with a rather all-inclusive checklist for your use in determining compliance. (See Figure 15-11.)

Self-Inspection Checklist

Flammable and Combustible Liquids

_____ Are combustible scrap, debris, and waste materials (oily rags, etc.) stored in covered metal receptacles and removed from the worksite promptly?

_____ Is proper storage practiced to minimize the risk of fire, including spontaneous combustion?

_____ Are approved containers and tanks used for the storage and handling of flammable and combustible liquids?

_____ Are all connections on drums and combustible liquid piping vapor and liquid tight?

_____ Are all flammable liquids kept in closed containers when not in use (for example, parts cleaning tanks, pans, etc.)?

_____ Are bulk drums of flammable liquids grounded and bonded to containers during dispensing?

_____ Do storage rooms for flammable and combustible liquids have explosion-proof lights?

_____ Do storage rooms for flammable and combustible liquids have mechanical or gravity ventilation?

_____ Is liquefied petroleum gas stored, handled, and used in accordance with safe practices and standards?

_____ Are "NO SMOKING" signs posted on liquefied petroleum gas tanks?

_____ Are liquefied petroleum storage tanks guarded to prevent damage from vehicles?

_____ Are all solvent wastes and flammable liquids kept in fire-resistant, covered containers until they are removed from the worksite?

_____ Is vacuuming used whenever possible rather than blowing or sweeping combustible dust?

_____ Are firm separators placed between containers of combustibles or flammables, when stacked one upon another, to assure their support and stability?

_____ Are fuel gas cylinders and oxygen cylinders separated by distance, and fire-resistant barriers, while in storage?

_____ Are fire extinguishers selected and provided for the types of materials in areas where they are to be used?

_____ Class A: ordinary combustible material fires.

_____ Class B: flammable liquid, gas or grease fires.

_____ Class C: energized-electrical equipment fires.

_____ Are appropriate fire extinguishers mounted within 75 feet of outside areas containing flammable liquids, and within ten feet of any inside storage area for such materials?

_____ Are extinguishers free from obstructions or blockage?

_____ Are all extinguishers serviced, maintained and tagged at intervals not to exceed one year?

_____ Are all extinguishers fully charged and in their designated places?

Figure 15-11. Flammable and combustible liquids. Courtesy of Occupational Safety and Health Administration

	Where sprinkler systems are permanently installed, are the nozzle heads so directed or arranged that water will not be sprayed into operating electrical switch boards and equipment?
_____	Are "NO SMOKING" signs posted where appropriate in areas where flammable or combustible materials are used or stored?
_____	Are safety cans used for dispensing flammable or combustible liquids at a point of use?
_____	Are all spills of flammable or combustible liquids cleaned up promptly?
_____	Are storage tanks adequately vented to prevent the development of excessive vacuum or pressure as a result of filling, emptying, or atmosphere temperature changes?
_____	Are storage tanks equipped with emergency venting that will relieve excessive internal pressure caused by fire exposure?
_____	Are "NO SMOKING" rules enforced in areas involving storage and use of hazardous materials?

Figure 15-11. Flammable and combustible liquid checklist. Courtesy of the Occupational Safety and Health Administration (Continued)

COMPRESSED GAS SAFETY

General Requirements (29 CFR 1910.101)

Cylinder Inspection

Employers shall determine that compressed gas cylinders under their control are in a safe condition to the extent that this can be determined by visual inspection. Visual and "other" inspections are required, but "other" inspections are not defined. These inspections must be conducted as prescribed in the Hazardous Materials Regulations of the Department of Transportation (DOT) contained in 49 CFR Parts 171-179 and 14 CFR Part 103. Where these regulations are not applicable, these inspections shall be conducted in accordance with Compressed Gas Association (CGA) Pamphlets C-6 and C-8. According to DOT regulations:

1. Cylinder that leaks, is bulged, has defective valves or safety devices; bears evidence of physical abuse, fire or heat damage, or detrimental rusting or corrosion must not be used unless it is properly repaired and requalified as prescribed in these regulations."

2. The term "cylinder" is defined as a pressure vessel designed for pressures higher than 40 pounds per square inch absolute and having a circular cross section. It does not include a portable tank, multiunit tank, car tank, cargo tank, or tanker car.

DOT requires the marking of basic information on all cylinders. Each required marking on a cylinder must be maintained so that it is legible. A summary of the DOT marking requirements is shown below:

1. DOT or ICC markings may appear—new manufacture must read "DOT," indicating compliance with 49 CFR 171.14. "3AA" indicates specifications in 49 CFR 178.37. "2015" is the marked service pressure.

2. Serial number—no duplications permitted with any particular symbol-serial number combinations.

3. Symbol of manufacturer, user, or purchaser.

4. "6 56," date of manufacture. Month and year. "0" disinterested inspector's official mark.

5. Plus mark (+) indicates cylinder may be ten percent overcharged per 49 CFR 173.302(c).

6. Retest dates.

7. Five-pointed star indicates ten-year retest interval. Reference 49 CFR 173.34(e)(15) for more information. (See Figure 15-12.)

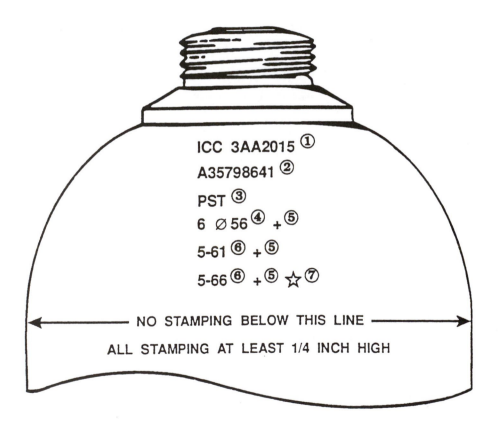

Figure 15-12. Marking on compressed gas cylinders. Courtesy of the Occupational Safety and Health Administration

Handling, Storage, and Utilization

The in-plant handling, storage, and utilization of all compressed gases in cylinders, portable tanks, railroad tank cars, or motor vehicle cargo tanks shall be in accordance with Compressed Gas Association (CGA) Pamphlet P-1.

Safety Relief Devices

Compressed gas cylinders, portable tanks, and cargo tanks shall have pressure relief devices installed and maintained in accordance with CGA Pamphlets S-1.1 and S-1.2.

Specific Gases

The OSHA regulations contain some sections regulating specific compressed gases, including acetylene, hydrogen, oxygen, nitrous oxide, anhydrous ammonia, and liquefied petroleum gases. The following shows where these regulations are found:

- 29 CFR 1910.102—Acetylene
- 29 CFR 1910.103—Hydrogen
- 29 CFR 1910.104—Oxygen
- 29 CFR 1910.105—Nitrous Oxide
- 29 CFR 1910.110—Liquefied Petroleum Gas
- 29 CFR 1910.111—Anhydrous Ammonia

There are many compressed gases in common use that fall under regulation 1910.101. Examples include chlorine, vinyl chloride, sulfur dioxide, methyl chloride, hydrogen sulfide, ethane, compressed air, and nitrogen. These compressed gases do not receive explicit coverage by the OSHA regulations, but are covered by the requirements of 1910.101.

Oxygen Fuel Gas Welding and Cutting (29 CFR 1910.253)

Flammable mixture

Mixtures of fuel gases and air or oxygen may be explosive and shall be guarded against. No device or attachment facilitating or permitting mixtures of air or oxygen with flammable gases prior to consumption, except at the burner or in a standard torch, shall be allowed unless approved for the purpose.

Maximum pressure

Under no condition shall acetylene be generated, piped (except approved cylinder manifolds) or utilized at a pressure in excess of 15 pounds per square inch gauge (103 kPa gauge pressure) or 30 pounds per square inch gauge (206 kPa absolute). (The 30 pounds per square inch gauge (206 kPa absolute) limit is intended to prevent unsafe use of acetylene in pressurized chambers such as caissons, underground excavations, or tunnel construction.) This requirement is not intended to apply to storage of acetylene dissolved in a suitable solvent in cylinders manufactured and maintained according to U.S. DOT requirements, or to acetylene for chemical use. The use of liquid acetylene shall be prohibited.

Cylinders and Containers

Approval and Marking

All portable cylinders used for the storage and shipment of compressed gases shall be constructed and maintained in accordance with regulations of the U.S. DOT, 49 CFR parts 171-179.

Compressed gas cylinders shall be legibly marked, for the purpose of identifying the gas content, with either the chemical or trade name of the gas. Such marking shall be by means of stenciling, stamping, or labeling, and shall not be readily removable. Whenever practical, the marking shall be located on the shoulder of the cylinder.

Storage of Cylinders—General

Cylinders shall be kept away from radiators and other sources of heat. Inside buildings, cylinders shall be stored in a well-protected, well-ventilated, dry location, at least 20 feet (6.1 meters) from highly combustible materials. Cylinders should be stored in definitely assigned places away from elevators, stairs, or gangways, or other areas where they might be knocked over or damaged by passing or falling objects, or subject to tampering.

Empty cylinders shall have their valves closed. Valve protection caps, where the cylinder is designed to accept a cap, shall always be in place, hand-tight, except when cylinders are in use or connected for use. The valve protection cap is designed to take the blow in case the cylinder falls.

Fuel Gas Cylinder Storage

Inside a building, cylinders, except those in actual use or attached ready for use, shall be limited to a total gas capacity of 2,000 cubic feet (56 cubic meters) or 300 pounds of liquefied petroleum gas.

Acetylene cylinders shall be stored valve end up. If the cylinder is on its side, acetone may leak out and create a dangerous condition.

Oxygen Storage

Oxygen cylinders in storage shall be separated from fuel-gas cylinders or combustible materials (especially oil or grease) a minimum distance of 20 feet (6.1 meters) or by a noncombustible barrier at least five feet (1.5 meters) high, having a fire-resistance rating of at least one-half hour. This requirement is intended to reduce the possibility of oxygen supporting any fire which occurs among the fuel gas storage.

Operating Procedures

Cylinders, cylinder valves, couplings, regulators, hose and apparatus shall be kept free from oily and greasy substances. Oxygen cylinders and apparatus shall not be handled with oily hands or gloves. A jet of oxygen must never be permitted to strike an oily surface, greasy clothes, or enter a fuel oil or other storage tank.

Valve-protection caps shall not be used for lifting cylinders from one vertical position to another. The cap may accidentally and suddenly come loose. Should the cylinder fall or be knocked over, the valve may be damaged or sheared off, causing a sudden release of pressure.

Should the valve outlet of a cylinder become clogged with ice, thaw the valve with warm (not boiling) water. Unless cylinders are secured on a special truck, regulators shall be removed and valve-protection caps, when provided for, shall be put in place before the cylinders are moved.

Cylinders not having fixed hand-wheels shall have keys, handles, or non-adjustable wrenches on valve stems while these cylinders are in service. Unless connected to a manifold, always attach a regulator to the compressed gas cylinder before use. Make certain that the regulator is proper for the particular gas and service pressure. Make sure the regulator is clean and has a clean filter installed in its inlet nipple.

Before attaching the regulator, remove the protective cap from the cylinder. Stand to one side of the cylinder. Open the cylinder valve slightly for an instant, and then close it. This "cracking" of the cylinder valve will clean the valve of dust or dirt, which may have accumulated during storage. Dirt can damage critical parts of a regulator, and may cause a fire or explosion. Before a regulator is removed from a cylinder valve, the valve shall be closed.

Compressed gas cylinders can present a variety of hazards due to their pressure and/or contents. General handling and storage requirements apply to all use of compressed gases. The applications of compressed gases are as follows:

- Use of inert gases—Gases that are non-flammable and non-toxic, but which may cause asphyxiation due to displacement of oxygen in poorly ventilated spaces.

- Use of flammable, low toxicity—Gases which are flammable (at a concentration in air of 13 percent by volume or have a flammable range wider than 13 percent by volume), but act as non-toxic, simple asphyxiants (e.g. hydrogen and methane).

- Use of pyrophoric gases and liquids—Gases or liquids, which spontaneously ignite on contact with air at a temperature of 130 degrees Fahrenheit or below.

- Use of corrosive, toxic, and highly toxic gases—Gases which may cause acute or chronic health effects at relatively low concentrations in air.

- Use of compressed gases in fume hoods.

- Use of cryogenic liquids—Use of liquids with a boiling point below -238 degrees Fahrenheit (-150 degrees Celsius).

- Fuel gases for welding, cutting, brazing—Use of oxygen and fuel gases (e.g. propane and acetylene) for gas welding and cutting applications.

Use of Fuel Gases for Welding and Cutting

OSHA lists requirements for oxygen-fuel gas welding and cutting in 29 CFR 1910.253. Cylinder handling precautions, materials of construction, and additional requirements are listed. This information should be reviewed by persons who will be using acetylene, oxygen, and other fuel gases and those who are designing facilities and equipment for this purpose. Be sure that all fuel gases are shut off at the cylinder valve after each use.

Compressed Gas Cylinders (29 CFR 1926.350)

Compressed gas cylinders have the potential to become a guided missile, if broken or damaged, and will release a tremendous amount of energy. Thus, compressed gas cylinders should be treated with respect. Most compressed gas cylinders are approximately 1/4 inch thick, weighing 150 pounds or more, and are under some 2,200 pounds per square inch of pressure.

When transporting, moving, or storing compressed gas cylinders, valve protection caps should be in place and secured. When cylinders are hoisted, they must be secured on a cradle, slingboard, or pallet (see Figure 15-13 for an example of a compressed gas cylinder lifting cradle). They should never be hoisted or transported by means of magnets or choker slings. Valve protection caps are never used for lifting cylinders from one vertical position to another. Bars should not be used under valves or valve protection caps to pry cylinders loose when frozen. Warm, not boiling, water is used to thaw cylinders loose.

*Figure 15-13. Cradle for lifting
compressed gas cylinders*

*Figure 15-14. Secured compressed gas
cylinders*

Compressed gas cylinders can be moved by tilting and rolling them on their bottom edges, but care must be taken to not drop or strike them together. This is especially true while transporting them by powered vehicles, since they must be secured in a vertical position. In most cases, regulators are removed, and valve protection caps put in place, before cylinders are moved. (See Figure 15-14, which shows cylinders that have been secured against falling.)

Oxygen cylinders which are put in storage are to be separated from fuel-gas cylinders or combustible materials (especially oil or grease). They must be separated by a minimum distance of 20 feet (6.1 meters), or by a noncombustible barrier of at least five feet (1.5 meters) high; this barrier must have a fire-resistance rating of at least one-half hour. Cylinders should be stored in assigned places which are away from elevators, stairs, and gangways; in areas where they will not be knocked over or damaged by passing or falling objects; or in areas where they are not subject to tampering by unauthorized persons.

Cylinders are to be kept far enough away from the actual welding or cutting operation so that sparks, hot slag, and flame will not reach them. When this is impractical, fire-resistant shields must be provided.

Also, cylinders should be placed where they cannot become part of an electrical circuit. Cylinders containing oxygen, acetylene, or other fuel gas must not be taken into confined spaces. Cylinders, whether full or empty, are not to be used as rollers or supports. No damaged or defective cylinders are to be used.

All employees should be trained in the safe use of fuel gas. The training should include how to crack a cylinder valve to clear dust or dirt prior to connecting a regulator. The worker cracking the valve should stand to the side and make sure the fuel gas is not close to an ignition source. The cylinder valve should always be opened slowly to prevent damage to the regulator. For quick closing, valves on fuel gas cylinders should not be opened more than one and a half turns. When a special wrench is required, it must be left in position on the stem of the valve. It must be kept in this position while the cylinder is in use, so that the fuel gas flow can be shut off quickly in the event of an emergency. In the case of manifolded or coupled cylinders, at least one such wrench shall always be available for immediate use. Nothing may be placed on top of a fuel gas cylinder when it is in use; this may damage the safety device or interfere with the quick closing of the valve.

Fuel gas shall not be used from cylinders unless a suitable regulator is attached to the cylinder valve or manifold. Before a regulator is removed from a cylinder valve, the cylinder valve must always be closed and the gas released from the regulator. If and when the valve on a fuel gas cylinder is opened, and there is found to be a leak around the valve stem, the valve shall be closed and the gland nut tightened. If this action does not stop the leak, the use of the cylinder must be discontinued, and it shall be properly tagged and removed from the work area. In the event that fuel gas should leak from the cylinder valve, rather than from the valve stem, and the gas cannot be shut off, the cylinder shall be properly tagged and removed from the work area. If a regulator attached to a cylinder valve will effectively stop a leak through the valve seat, the cylinder need not be removed from the work area. But, if a leak should develop at a fuse plug or other safety device, the cylinder must be removed from the work area.

Cryogenic Liquids

Cryogenic liquids, as with all gases, must be ordered through proper channels. Only inert gases should be permitted in portable cryogenic containers. Liquid oxygen, liquid hydrogen and other flammable or toxic cryogenic liquids should not be used. Where appropriate, exterior tanks of liquid oxygen used as a source of gaseous oxygen may be installed, when appropriate clearances and safety precautions are followed.

All cryogenic liquids should be used with caution due to the potential for skin or eye damage resulting from low temperature, and hazards associated with pressure buildups in enclosed piping and containers. Portable containers should only be used where there is sufficient ventilation. Do not place containers in a closet or other enclosed space where there is no ventilation supply to the area. The buildup of inert gas in such an area can generate an oxygen deficient atmosphere.

A full face shield, loose fitting cryogenic handling gloves, apron, and cuffless slacks are the recommended equipment for transferring cryogenic fluids. Special vacuum jacket containers with loose fitting lids should be used to handle small quantities. Vacuum jacketed containers, provided by the gas supplier, will have overpressure relief devices in place. When plumbing cryogenic liquids, it is very important to include a pressure relief valve between any two shutoff valves. Also, any space where cryogenic fluids may accumulate (consider leakage into enclosed equipment as well) must be protected by overpressure relief devices.

Tremendous pressures can be obtained in enclosed spaces as the liquid converts to gas. For example, one cubic centimeter of liquid nitrogen will expand to 700 times this volume as it converts (warms) to its gaseous state. Lines carrying liquid should be well insulated.

Containers to be filled with cryogenic liquids should be filled slowly to avoid splashing. Cryogenic containers showing evidence of loss of vacuum in their outer jacket (ice buildup on the outside of the container) should not be accepted from the gas supplier, and they should be removed from service. Contact with air (or gases with a higher boiling point) can cause an ice plug in a cryogenic container. Should ice plugs be noted, contact knowledgeable safety officials.

Summary

The use of compressed gases can be accomplished if safe handling and storage procedures, which are found in the OSHA regulations, are followed. Respect for the potential power which can be released from compressed cylinders, should be an integral part of all training programs. Good training programs should also teach the identification and proper response to hazards of compressed gas.

REFERENCES

Reese, C. D. and J.V. Eidson. *Handbook of OSHA Construction Safety & Health.* Boca Raton, FL: CRC/Lewis Publishers, 1999.

United States Department of Labor. Occupational Safety and Health Administration. *OSHA 10 and 30 Hour Construction Safety and Health Outreach Training Manual.* Washington, DC, 1991.

United States Department of Labor. Occupational Safety and Health Administration. General Industry. *Code of Federal Regulations.* Title 29, Part 1910. Washington, GPO, 1998.

United States Department of Labor. Occupational Safety and Health Administration. Construction. *Code of Federal Regulations.* Title 29, Part 1926. Washington, GPO, 1998.

United States Department of Labor. Occupational Safety and Health Administration. Office of Training and Education. *OSHA Voluntary Compliance Outreach Program: Instructors Reference Manual.* Des Plaines, IL, 1993.

Unites States Department of Labor. Occupational Safety and Health Administration. OSHA Web Site for Small Business. *Flammable and Combustible Liquids - 1910.106. Washington,* DC, 1997.

CHAPTER 16

SUMMARY

Use of a vacuum lift, which requires a very small effort to lift and position boxes

As you look, everywhere you travel, you will see materials being handled, lifted, or moved. Most of us probably have no true concept as to how pervasive material handling activities are within our society today. Just think for a minute: whether we are talking about material handling in the form of garbage, mail, air freight, movement by water, or railroad, there is a myriad of devices designed to accomplish the least to the most sophisticated material handling tasks. In summary, it would seem appropriate to give a pictorial presentation of the variety of activities, as well as devices and vehicles, which are in use for material handling activities.

MANUAL LIFTING

Manual lifting will not go away. Thus, whenever a manual lifting task is performed, the safe approach to accomplishing it should be given careful and due consideration. In these cases, where an individual is expected to carry and move loads, the risk are definitely heightened. (See Figure 16-1.)

Figure 16-1. Manual handling of materials

POWER OPERATED LIFTING DEVICES

Any time power devices are used for lifting and moving loads, the power alone of these devices makes for a hazardous situation. This is why operator training, safe operating procedures, maintenance and attention to potential hazards is stressed, since material handling aids are only effective if the task can be done safely and free from hazardous incidents. (See Figure 16-2.)

Figure 16-2. Powered material handling activity using a truck mounted crane and wire rope slings to lift a concrete block

SPECIAL LIFTING VEHICLES

Many special lifting devices have evolved over a period of years. There are usually such devices available for purchase, which can meet your specific material handling needs. (See Figure 16-3.)

Figure 16-3. Special handling device such as this automated turntable for shrink wrapping a pallet

TRUCKS FOR MATERIAL HANDLING

Many types of vehicles with a wide variety of lifting adaptations have been introduced. Delivery vehicles with their own cranes, lift gates, and other handling devices are commonplace today. The presence of ramp access for vehicles is widely available, as well as all sizes and shapes of vehicles to meet special material handling tasks. (See Figures 16-4 and 16-5.)

Figure 16-4. Delivery vehicle with its own forklift for loading and unloading

Figure 16-5. A stock picker is an example of a special use vehicle

MORE MATERIAL HANDLING

What you have seen in this book only scratches the surface regarding the devices for improving materials handling activities so that materials handling can be accomplished in a safe manner. The following are only pictorial examples of some of the many types of material handling equipment.

Dump truck

Concrete truck

Small handling equipment commonly referred to as a "bobcat"

Large off-road truck

The versatile backhoe

Use of chain sling to move concrete

A truck-mounted crane

A highway tandem trailer rig

Use of pipe jacks to support pipe being worked on

The "old faithful" car jack

The multipurpose platform truck

The floor crane

The use of palletized loads and pallet trucks

A well-designed material handling workstation

A stationary hydraulic-lift work platform makes accessing stock from carousel stock bins easy

These are only a few of the many pieces of equipment and devices which are in use today to assist with material handling activities. The effective use of these and other such devices will go a long way toward the improving safety and decreasing injuries related to material handling tasks.

APPENDIX A

OSHA TRAINING REQUIREMENT FOR MATERIAL HANDLING ACTIVITIES

General Industry Training Requirements from 29 CFR 1910:

Flammable and Combustible Liquids 1910.106(b)(5)(v)(2) and (3)	(2)	That detailed printed instructions of what to do in flood emergencies are properly posted.
	(3)	That station operators and other employees depended upon to carry out such instructions are thoroughly informed as to the location and operation of such valves and other equipment necessary to effect these requirements.
Explosive and Blasting Agents 1910.109(d)(3)(i) and (iii)	(i)	Vehicles transporting explosives shall only be driven by and be in the charge of a driver who is familiar with the traffic regulations, State laws, and the provisions of this section.
	(iii)	Every motor vehicle transporting any quantity of Class A or Class B explosives shall, at all times, be attended by a driver or other attendant of the motor carrier. This attendant shall have been made aware of the class of the explosive material in the vehicle and of its inherent dangers, and shall have been instructed in the measures and procedures to be followed in order to protect themselves from those dangers. He shall have been made familiar with the vehicle he is assigned, and shall be trained, supplied with the necessary means, and authorized to move the vehicle when required.
1910.109(g)(3)(iii)(a)	(iii)(a)	The operator shall be trained in the safe operation of the vehicle together with its mixing, conveying, and related equipment. The employer shall assure that the operator is familiar with the commodities being delivered and the general procedure for handling emergency situations.

1910.109(g)(6)(ii)   (ii)   Vehicles transporting blasting agents shall only be driven by and be in charge of a driver in possession of a valid motor vehicle operator's license. Such a person shall also be familiar with the State's vehicle and traffic laws.

Bulk Delivery and   (iii)   The operator shall be trained in the safe operation of the vehicle together with its mixing, conveying, and related equipment. He shall be familiar with the commodities being delivered and the general procedure for handling emergency situations.
Mixing Vehicles
1910.109(h)(3)(d)(iii)

Storage and Handling of   (16)   Instructions. Personnel performing installation, removal, operation, and maintenance work shall be properly trained in such functions.
Liquefied Petroleum
Gases 1910.110(b)(16)

1910.110(d)(12)(i)   (i)   When standard watch service is provided, it shall be extended to the LP-Gas installation and personnel properly trained.

1910.111(b)(13)(ii)   (ii)   The employer shall insure that unloading operations are performed by reliable persons properly instructed and given the authority to monitor careful compliance with all applicable procedures.

Specifications for   (1)(ii)   All employees shall be instructed that danger signs indicate immediate danger and that special precautions are necessary.
Accident Prevention
Signs and Tags
 1910.145(c)(1)(ii),

1910.145(c)(2)(ii)   (2)(ii)   All employees shall be instructed that caution signs indicate a possible hazard against which proper precautions should be taken.

1910.145(c)(3)   (3)   Safety instruction signs. Safety instruction signs shall be used where there is a need for general instructions and suggestions relative to safety measures.

Powered Industrial Trucks   (1)   Operator training. Only trained and authorized operators shall be permitted to operate a powered industrial truck. Methods shall be devised to train operators in the safe operation of powered industrial trucks.
1910.178(1)

1910.178(l)(1)(i)   (i)   Safe operation. The employer shall ensure that each powered industrial truck operator is competent to operate a powered industrial truck safely, as demonstrated by the successful completion of the training and evaluation specified in this paragraph (l).

1910.178(l)(1)(ii)	(ii)	Prior to permitting an employee to operate a powered industrial truck (except for training purposes), the employer shall ensure that each operator has successfully completed the training required by this paragraph (l), except as permitted by paragraph (l)(5).
1910.178(l)(2)(i)	(i)	Training program implementation. Trainees may operate a powered industrial truck only:
1910.178(l)(2)(i)(A)	(A)	Under the direct supervision of persons who have the knowledge, training, and experience to train operators and evaluate their competence; and
1910.178(l)(2)(i)(B)	(B)	Where such operation does not endanger the trainee or other employees.
1910.178(l)(2)(ii)	(ii)	Training shall consist of a combination of formal instruction (e.g., lecture, discussion, interactive computer learning, video tape, written material), practical training (demonstrations performed by the trainer and practical exercises performed by the trainee), and evaluation of the operator's performance in the workplace.
1910.178(l)(2)(iii)	(iii)	All operator training and evaluation shall be conducted by persons who have the knowledge, training, and experience to train powered industrial truck operators and evaluate their competence.
1910.178(l)(3)	(3)	Training program content. Powered industrial truck operators shall receive initial training in the following topics, except in topics which the employer can demonstrate are not applicable to safe operation of the truck in the employer's workplace.
1910.178(l)(3)(i)	(i)	Truck-related topics:
1910.178(l)(3)(i)(A)	(A)	Operating instructions, warnings, and precautions for the types of truck the operator will be authorized to operate;
1910.178(l)(3)(i)(B)	(B)	Differences between the truck and the automobile;
1910.178(l)(3)(i)(C)	(C)	Truck controls and instrumentation: where they are located, what they do, and how they work;
1910.178(l)(3)(i)(D)	(D)	Engine or motor operation;
1910.178(l)(3)(i)(E)	(E)	Steering and maneuvering;

1910.178(l)(3)(i)(F)	(F)	Visibility (including restrictions due to loading);
1910.178(l)(3)(i)(G)	(G)	Fork and attachment adaptation, operation, and use limitations;
1910.178(l)(3)(i)(H)	(H)	Vehicle capacity;
1910.178(l)(3)(i)(I)	(I)	Vehicle stability;
1910.178(l)(3)(i)(J)	(J)	Any vehicle inspection and maintenance that the operator will be required to perform;
1910.178(l)(3)(i)(K)	(K)	Refueling and/or charging and recharging of batteries;
1910.178(l)(3)(i)(L)	(L)	Operating limitations;
1910.178(l)(3)(i)(M)	(M)	Any other operating instructions, warnings, or precautions listed in the operator's manual for the types of vehicle that the employee is being trained to operate.
1910.178(l)(3)(ii)	(ii)	Workplace-related topics:
1810.178(l)(3)(ii)(A)	(A)	Surface conditions where the vehicle will be operated;
1910.178(l)(3)(ii)(B)	(B)	Composition of loads to be carried and load stability;
1910.178(l)(3)(ii)(C)	(C)	Load manipulation, stacking, and unstacking;
1910.178(l)(3)(ii)(D)	(D)	Pedestrian traffic in areas where the vehicle will be operated;
1910.178(l)(3)(ii)(E)	(E)	Narrow aisles and other restricted places where the vehicle will be operated;
1910.178(l)(3)(ii)(F)	(F)	Hazardous (classified) locations where the vehicle will be operated;
1910.178(l)(3)(ii)(G)	(G)	Ramps and other sloped surfaces that could affect the vehicle's stability;
1910.178(l)(3)(ii)(H)	(H)	Closed environments and other areas where insufficient ventilation or poor vehicle maintenance could cause a buildup of carbon monoxide or diesel exhaust;
1910.178(l)(3)(ii)(I)	(I)	Other unique or potentially hazardous environmental conditions in the workplace that could affect safe operation.

1910.178(l)(3)(iii)	(iii)	The requirements of this section.
1910.178(l)(4)	(4)	Refresher training and evaluation.
1910.178(l)(4)(i)	(i)	Refresher training, including an evaluation of the effectiveness of that training, shall be conducted as required by paragraph (l)(4)(ii) to ensure that the operator has the knowledge and skills needed to operate the powered industrial truck safely.
1910.178(l)(4)(ii)	(ii)	Refresher training in relevant topics shall be provided to the operator when:
19810.178(l)(4)(ii)(A)	(A)	The operator has been observed to operate the vehicle in an unsafe manner;
1910.178(l)(4)(ii)(B)	(B)	The operator has been involved in an accident or near-miss incident;
1910.178(l)(4)(ii)(C)	(C)	The operator has received an evaluation that reveals that the operator is not operating the truck safely;
1910.178(l)(4)(ii)(D)	(D)	The operator is assigned to drive a different type of truck; or
1910.178(l)(4)(ii)(E)	(E)	A condition in the workplace changes in a manner that could affect safe operation of the truck.
1910.178(l)(4)(iii)	(iii)	An evaluation of each powered industrial truck operator's performance shall be conducted at least once every three years.
1910.178(l)(5)	(5)	Avoidance of duplicative training. If an operator has previously received training in a topic specified in paragraph (l)(3) of this section, and such training is appropriate to the truck and working conditions encountered, additional training in that topic is not required if the operator has been evaluated and found competent to operate the truck safely.
1910.178(l)(6)	(6)	Certification. The employer shall certify that each operator has been trained and evaluated as required by this paragraph (l). The certification shall include the name of the operator, the date of the training, the date of the evaluation, and the identity of the person(s) performing the training or evaluation.
1910.178(l)(7)	(7)	Dates. The employer shall ensure that operators of powered industrial trucks are trained, as appropriate, by the dates shown in the following table.

If the employee was hired:	The initial training and evaluation of that must be completed:
Before December 1, 1999	By December 1, 1999.
After December 1, 1999	Before the employee is assigned to operate a powered industrial truck.

1910.178(l)(8)	(8)	Appendix A to this section provides non-mandatory guidance to assist employers in implementing this paragraph (l). This appendix does not add to, alter, or reduce the requirements of this section.
Moving the Load 1910.179(n)(3)(ix)	(ix)	When two or more cranes are used to lift a load, one qualified responsible person shall be in charge of the operation. He shall analyze the operation and instruct all personnel involved in the proper positioning, rigging of the load, and the movements to be made.
1910.179(a)(3)	(3)	Fire extinguishers. The employer shall insure that operators are familiar with the operation and care of fire extinguishers provided.
Crawler Locomotives and Truck Cranes 1910.180(i)(5)(ii)	(ii)	Operating and maintenance personnel shall be made familiar with the use and care of the fire extinguishers provided.
Grain Handling Facilities 1910.272(e)(1)(i) and (2)	(e)	Training. (1) The employer shall provide training to employees at least annually and when changes in job assignment will expose them to new hazards.
		Current employees, and new employees prior to starting work, shall be trained in at least the following:
	(ii)	Specific procedures and safety practices applicable to their job tasks including but not limited to, cleaning procedures for grinding equipment, clearing procedures for choked legs, housekeeping procedures, hot work procedures, preventive maintenance procedures and lock-out/tag-out procedures.
	(2)	Employees assigned special tasks, such as bin entry and handling of flammable or toxic substances, shall be provided training to perform these tasks safely.

Shipyard Employment Training Requirements from 29 CFR 1915:

Ropes, Chains and Slings 1915.112(c)(5)	(5)	All repairs to chains shall be made under qualified supervision. Links or portions of the chain found to be defective as described in paragraph (4) of this section shall be replaced by links having proper dimensions and made of material similar to that of the chain. Before repaired chains are returned to service, they shall be proof tested to the proof test load recommended by the manufacturer.
Use of Gear 1915.116(1)	(1)	An individual who is familiar with the signal code in use shall be assigned to act as a signalman when the hoist operator cannot see the load being handled. Communications shall be made by means of clear and distinct visual or auditory signals except that verbal signals shall not be permitted.
Qualifications of Operators 1915.117(a) and (b)		Paragraph (a) of this section shall apply to ship repairing and shipbuilding only, and Paragraph (b) of this section shall apply to ship repairing, shipbuilding and shipbreaking.
	(a)	When ship's gear is used to hoist materials aboard, a competent person shall determine that the gear is properly rigged, that it is in safe condition, and that it will not be overloaded by the size and weight of the lift.
	(b)	Only those employees who understand the signs, notices, and operating instructions, and are familiar with the signal code in use, shall be permitted to operate a crane, winch, or other power operated hoisting apparatus.

Construction Training Requirements from 29 CFR 1926:

General Safety and Health Provisions 1926.20(b(2) and (4))	(2)	Such programs [as may be necessary to comply with this part] shall provide for frequent and regular inspections of the job sites, materials, and equipment to be made by competent persons [capable of identifying existing and predictable hazards in the surroundings or working conditions which are unsanitary, hazardous, or dangerous to employees, and who have authorization to take prompt corrective measures to eliminate them designated by the employers].

| | (4) | The employer shall permit only those employees qualified [one who, by possession of a recognized degree, certificate, or professional standing, or who by extensive knowledge, training, and experience, has successfully demonstrated his ability to solve or resolve problems relating to the subject matter, the work, or the project] by training or experience to operate equipment and machinery. |

Safety Training and Education 1926.21(a)

(a) General requirements. The Secretary shall, pursuant to section (7)(f) of the Act, establish and supervise programs for the education and training of employers and employees in the recognition and prevention of unsafe conditions in employments covered by the Act.

1926.21(b)(1) through (3)

(1) The employer should avail himself of the safety and health training programs the Secretary provides.

(2) The employer shall instruct each employee in the recognition and avoidance of unsafe conditions and the regulations applicable to his work environment to control or eliminate any hazards or other exposure to illness or injury.

(3) Employees required to handle or use poisons, caustics, and other harmful substances shall be instructed regarding their safe handling and use, and be made aware of the potential hazards, personal hygiene, and personal protective measures required.

1926.21(b)(5)

(5) Employees required to handle or use flammable liquids, gases, or toxic materials shall be instructed in the safe handling and use of these materials and made aware of the specific requirements contained in Subpart, F, and other applicable subparts of this part.

Cranes and Derricks 1926.550(a)(1), (5) and (6)

(1) The employer shall comply with the manufacturer's specifications and limitations applicable to the operation of any and all cranes and derricks. Where manufacturer's specifications are not available, the limitations assigned to the equipment shall be based on the determinations of a qualified engineer competent in this field and such determinations will be appropriately documented and recorded. Attachments used with cranes shall not exceed the capacity, rating, or scope recommended by the manufacturer.

(5)   The employer shall designate a competent person who shall inspect all machinery and equipment prior to each use, and during use, to make sure it is in safe operating condition. Any deficiencies shall be repaired, or defective parts replaced, before continued use.

(6)   A thorough, annual inspection of the hoisting machinery shall be made by a competent person or by a government or private agency recognized by the U.S. Department of Labor. The employer shall maintain a record of the dates and results of inspections for each hoisting machine and piece of equipment.

Material Hoists,
Personnel Hoists, and
Elevators
1926.552(a)(1)

(1)   The employer shall comply with the manufacturer's specifications and limitations applicable to the operation of all hoists and elevators. Where manufacturer's specifications are not available, the limitations assigned to the equipment shall be based on the determinations of a professional engineer competent in the field.

Material Handling
Equipment
1926.602(c)(1)(vi)

(c)   Lifting and hauling equipment (other than equipment covered under Subpart N of this part).

(1)(vi) All industrial trucks in use shall meet the applicable requirements of design, construction, stability, inspection, testing, maintenance, and operation, as defined in American National Standards Institute B56.1-1969, Safety Standards for Powered Industrial Trucks.

From ANSI Standard B56.1-1969: Operator Training. "Only trained and authorized operators shall be permitted to operate a powered industrial truck. Methods shall be devised to train operators in the safe operation of powered industrial trucks. Badges or other visual indication of the operators' authorization should be displayed at all times during the work period."

APPENDIX B

HAZARD ASSESSMENT AND SELECTION
OF
PERSONAL PROTECTIVE EQUIPMENT

Courtesy of the Occupational Safety and Health Administration

Hazard Assessment Certification Form

Date:	Location:
Assessment Conducted By:	
Specific Tasks Performed at this Location:	

Hazard Assessment and Selection of Personal Protective Equipment

I. Overhead Hazards -
- Hazards to consider include:
- Suspended loads that could fall
- Overhead beams or loads that could be hit against
- Energized wires or equipment that could be hit against
- Employees work at elevated site who could drop objects on others below
- Sharp objects or corners at head level

Hazards Identified:

Head Protection

Hard Hat:	Yes	No
If yes, type:		

      ☐ **Type A** (impact and penetration resistance, plus low-voltage electrical insulation)

      ☐ **Type B** (impact and penetration resistance, plus high-voltage electrical insulation)

      ☐ **Type C** (impact and penetration resistance)

II. Eye and Face Hazards -
- Hazards to consider include:
- Chemical splashes
- Dust
- Smoke and fumes
- Welding operations
- Lasers/optical radiation
- Projectiles

Hazards Identified:

Eye Protection

Safety glasses or goggles	Yes	No
Face shield	Yes	No

III. Hand Hazards -
- Hazards to consider include:
- Chemicals
- Sharp edges, splinters, etc.
- Temperature extremes
- Biological agents
- Exposed electrical wires
- Sharp tools, machine parts, etc.
- Material handling

Hazards Identified:

Hand Protection

Gloves	Yes	No
☐ Chemical resistant ☐ Temperature resistant ☐ Abrasion resistant ☐ Other (Explain) ☐		

IV. Foot Hazards -
- Hazards to consider include:
- Heavy materials handled by employees
- Sharp edges or points (puncture risk)
- Exposed electrical wires
- Unusually slippery conditions
- Wet conditions
- Construction/demolition

Hazards Identified:

Foot Protection

Safety shoes	Yes	No
Types: ☐ Toe protection ☐ Metatarsal protection ☐ Puncture resistant ☐ Electrical insulation ☐ Other (Explain)		

V. Other Identified Safety and/or Health Hazards:

Hazard	Recommended Protection

I certify that the above inspection was performed to the best of my knowledge and ability, based on the hazards present on _____

_____
(Signature)

APPENDIX C

LOAD DOCK SAFETY CHECKLIST

Permission by the Rite-Hite Corporation

Loading Dock Safety Checklist

Provided as a service to industry by Rite-Hite Corporation.

Date _____

Company Name _____

Plant Name _____ Plant Location _____

Dock examined/Door numbers _____

Company representative completing checklist:

Name _____ Title _____

A. Vehicles/Traffic Control

1. Do forklifts have the following safety equipment?
   - ☐ Seat belt   ☐ Load backrest
   - ☐ Headlight   ☐ Backup alarm
   - ☐ Horn   ☐ Overhead guard
   - ☐ Tilt indicator
   - ☐ On-board fire extinguisher
   - ☐ Other _____

2. Are the following in use?

	Yes	No
Driver candidate screening	☐	☐
Driver training/licensing	☐	☐
Periodic driver retraining	☐	☐
Vehicle maintenance records	☐	☐
Written vehicle safety rules	☐	☐

	Yes	No
3. Is the dock kept clear of loads of materials?	☐	☐
4. Are there convex mirrors at blind corners?	☐	☐
5. Is forklift cross traffic over dock levelers restricted?	☐	☐
6. Is pedestrian traffic restricted in the dock area?	☐	☐
7. Is there a clearly marked pedestrian walkway?	☐	☐
8. Are guardrails used to define the pedestrian walkway?	☐	☐

9. Comments/recommendations

   _____
   _____
   _____
   _____
   _____

B. Vehicle Restraining

1. If vehicle restraints are used:

	Yes	No
a. Are all dock workers trained in the use of the restraints?	☐	☐
b. Are the restraints used consistently?	☐	☐
c. Are there warning signs and lights inside and out to tell when a trailer is secured and when it is not?	☐	☐
d. Are the outdoor signal lights clearly visible even in fog or bright sunlight?	☐	☐
e. Can the restraints secure trailers regardless of the height of their ICC bars?	☐	☐
f. Can the restraints secure all trailers with I-beam, round, or other common ICC bar shapes?	☐	☐
g. Does the restraint sound an alarm when a trailer cannot be secured because its ICC bar is missing or out of place?	☐	☐

RITE-HITE®

h. Are dock personnel specifi- ☐ ☐
cally trained to watch for
trailers with unusual rear-end
assemblies (e.g. sloping steel
back plates and hydraulic tailgates) that can cause a restraint
to give a signal that the trailer
is engaged even when it is not
safely engaged?

i. Are dock personnel specifi- ☐ ☐
cally trained to observe the
safe engagement of all unusual rear-end assemblies?

j. Do the restraints receive ☐ ☐
regular planned maintenance?

k. Do restraints need repairs or replacement?
(List door numbers.)

_____

l. Comments/recommendations

_____

_____

2. If wheel chocks are used: Yes No
a. Are dock workers, rather than ☐ ☐
truckers, responsible for
placing chocks?

b. Are all dock workers trained ☐ ☐
in proper chocking proce-
dures?

c. Are chocks of suitable design ☐ ☐
and construction?

d. Are there two chocks for each ☐ ☐
position?

e. Are all trailers chocked on ☐ ☐
both sides?

f. Are chocks chained to the ☐ ☐
building?

g. Are warning signs in use? ☐ ☐
h. Are driveways kept clear of ☐ ☐
ice and snow to help keep
chocks from slipping?

i. Comments/recommendations

_____

_____

. Dock Levelers and Ramps

Yes No
1. Are the dock levelers working ☐ ☐
properly?

2. Are levelers long enough to ☐ ☐
provide a gentle grade into
trailers of all heights?

3. Is leveler width adequate when ☐ ☐
servicing wide trailers?

4. Do platform width and con- ☐ ☐
figuration allow safe handling
of end loads?

5. Is leveler capacity adequate ☐ ☐
given typical load weights, lift
truck speeds, ramp inclines
and frequency of use?

6. Do levelers have the following
safety features?
Working-range toe guards ☐ ☐
Full-range toe guards ☐ ☐
Ramp free-fall protection ☐ ☐
Automatic recycling ☐ ☐

7. Do levelers receive regular
planned maintenance?

8. Do levelers need repair or replacement?
(List door numbers.)

_____

9. Are dock levelers capable of ☐ ☐
helping prevent accidental
falls of personnel or equipment from vacant loading
docks?

10. Are ramps kept clean and clear ☐ ☐
of debris, ice, snow or rain to
prevent slipping and sliding
during loading/unloading
operations?

11. Comments/recommendations _____

_____

_____

D. Portable Dock Plates

Yes No
1. Is plate length adequate? ☐ ☐
2. Is plate capacity adequate? ☐ ☐
3. Are plates of suitable design ☐ ☐
and materials?
4. Do plates have curbed sides? ☐ ☐
5. Do plates have suitable anchor ☐ ☐
stops?
6. Are plates moved by lift trucks ☐ ☐
rather than by hand?
7. Are plates stored away from ☐ ☐
traffic?
8. Are plates inspected regularly? ☐ ☐
9. Do plates need repair or replacement? (List
door numbers.)

_____

_____

10. Comments/recommendations

_____

_____

E. Dock Doors

	Yes	No
1. Are doors large enough to admit all loads without obstruction?	☐	☐
2. Are door rails protected by bumper posts?	☐	☐
3. Do doors receive regular planned maintenance?	☐	☐

4. Which doors (if any) need repair or replacement?

_____

5. Comments/recommendations

_____

_____

F. Traffic Doors

	Yes	No
1. Are doors wide enough to handle all loads and minimize damage?	☐	☐
2. Does door arrangement allow safe lift truck and pedestrian traffic?	☐	☐
3. Are visibility and lighting adequate on both sides of all doors?	☐	☐

4. Which doors (if any) need repair or replacement?

_____

5. Comments/recommendations

_____

_____

G. Weather Sealing

	Yes	No
1. Are the seals or shelters effective in excluding moisture and debris from the dock?	☐	☐
2. Are seals or shelters sized to provide an effective seal against all types of trailers?	☐	☐
3. Are seals or shelters designed so that they will not obstruct loading and unloading?	☐	☐

	Yes	No
4. Are dock levelers weather sealed along the sides and back?	☐	☐
5. In addition to seals or shelters, would an air curtain solve a problem?	☐	☐

6. Do seals or shelters need repair or replacement? (List door numbers.)

_____

7. Comments/recommendations

_____

_____

H. Trailer Lifting

1. How are low-bed trailers elevated for loading/unloading?
   ☐ Wheel risers   ☐ Concrete ramps
   ☐ Trailer-mounted jacks
   ☐ Truck levelers

	Yes	No
2. Do lifting devices provide adequate stability?	☐	☐
3. Are trailers secured with vehicle restraints when elevated?	☐	☐

4. Do lifting devices need repair or replacement? (List door numbers.)

_____

5. Comments/recommendations

_____

_____

I. Other Considerations

	Yes	No
1. Dock lights		
a. Is lighting adequate inside trailers?	☐	☐
b. Is the lift mechanism properly shielded?	☐	☐
2. Scissors lifts		
a. Are all appropriate workers trained in safe operating procedures?	☐	☐
b. Is the lift mechanism properly shielded?	☐	☐
c. Are guardrails and chock ramps in place and in good repair?	☐	☐

3. Conveyors
   a. Are all appropriate workers ☐ ☐
      trained in safe operating
      procedures?
   b. Are necessary safeguards in ☐ ☐
      place to protect against
      pinch points, jam-ups and
      runaway material?
   c. Are crossovers provided? ☐ ☐
   d. Are emergency stop buttons ☐ ☐
      in place and properly
      located?

4. Strapping
   a. Are proper tools available ☐ ☐
      for applying strapping?
   b. Do workers cut strapping ☐ ☐
      using only cutters equipped
      with a holddown device?
   c. Do workers wear hand, foot ☐ ☐
      and face protection when
      applying and cutting
      strapping?
   d. Are all appropriate workers ☐ ☐
      trained in safe strapping
      techniques?

5. Manual handling
   a. Is the dock designed so as to ☐ ☐
      minimize manual lifting and
      carrying?
   b. Are dock workers trained ☐ ☐
      in safe lifting and manual
      handling techniques?

6. Miscellaneous   Yes   No
   a. Are pallets regularly ☐ ☐
      inspected?
   b. Are dock bumpers in good ☐ ☐
      repair?
   c. Is the dock kept clean and ☐ ☐
      free of clutter?
   d. Are housekeeping inspec- ☐ ☐
      tions performed periodically?
   e. Are anti-skid floor surfaces, ☐ ☐
      mats or runners used where
      appropriate?

   f. Are stairways or ladders pro- ☐ ☐
      vided for access to ground
      level from the dock?
   g. Is the trailer landing strip in ☐ ☐
      good condition?
   h. Are dock approaches free of ☐ ☐
      potholes or deteriorated
      pavement?
   i. Are dock approaches and ☐ ☐
      outdoor stairs kept clear of
      ice and snow?
   j. Are dock positions marked ☐ ☐
      with lines or lights for accu-
      rate trailer spotting?
   k. Do all dock workers wear ☐ ☐
      personal protective equip-
      ment as required by com-
      pany policy?
   l. Is safety training provided ☐ ☐
      for all dock employees?
   m. Are periodice safety ☐ ☐
      refresher courses offered?

J. General Comments/Recommendations

_____
_____
_____
_____
_____
_____

For additional copies of this Loading Dock Safety Checklist,
write to Rite-Hite Corporation, 8900 Arbon Drive, P.O. Box
23043, Milwaukee, WI 53223-0043.
For more information on loading dock safety, write for a
complimentary copy of Rite-Hite's Dock Safety Guide.

This loading dock safety checklist is provided as a service by Rite-Hite
Corporation, Milwaukee. Wis. It is intended as an aid to safety evaluation
of loading dock equipment and operations. However, it is not intended
as a complete guide to loading dock hazard identification. Therefore,
Rite-Hite Corporation makes no guarantees as to nor assumes any
liability for the sufficiency or completeness of this document. It may be
necessary under particular circumstances to evaluate other dock equipment and procedures in addition to those included in the checklist.
For information on U.S. loading dock safety requirements, consult
OSHA Safety and Health Standards (29 CFR 1910). In other countries,
consult the applicable national or provincial occupational health and
safety codes.

Contact: Rite-Hite Corporation, 8900 N. Arbon Drive, P.O. Box 23043, Milwaukee, WI 53223
(414) 355-2600 ● 1-800-456-0600 ● FAX (414) 355-9248
Rite-Hite® is a registered trademark of the Rite-Hite Corporation.

GLOSSARY

CRAWLER LOCOMOTIVE AND TRUCK CRANES (29 CFR 1910.180):

Accessory—A secondary part or assembly of parts which contributes to the overall function and usefulness of a machine.

Angle indicator (boom)—An accessory which measures the angle of the boom to the horizontal.

ANSI—The American National Standards Institute.

Appointed—A person assigned specific responsibilities by the employer or the employer's representative.

Axis of rotation—The vertical axis around which the crane superstructure rotates.

Axle—The shaft or spindle with which or about which a wheel rotates. On truck- and wheel-mounted cranes it refers to an automotive type of axle assembly including housings, gearing, differential, bearings, and mounting appurtenances.

Axle (bogie)—Two or more automotive-type axles mounted in tandem in a frame so as to divide the load between the axles and permit vertical oscillation of the wheels.

Base (mounting)—The traveling base or carrier on which the rotating superstructure is mounted such as a car, truck, crawlers, or wheel platform.

Boom (crane)—A member hinged to the front of the rotating superstructure with the outer end supported by ropes leading to a gantry or A-frame and used for supporting the hoisting tackle.

Boom angle—The angle between the longitudinal centerline of the boom and the horizontal. The boom longitudinal centerline is a straight line between the boom foot pin (heel pin) centerline and boom point sheave pin centerline.

Boom hoist—A hoist drum and rope reeving system used to raise and lower the boom. The rope system may be all live reeving or a combination of live reeving and pendants.

Boom stop—A device used to limit the angle of the boom at the highest position.

Brake—A device used for retarding or stopping motion by friction or power means.

Cab—A housing which covers the rotating superstructure machinery and/or operator's station. On truck-crane trucks a separate cab covers the driver's station.

Clutch—A friction, electromagnetic, hydraulic, pneumatic, or positive mechanical device for engagement or disengagement of power.

Counterweight—A weight used to supplement the weight of the machine in providing stability for lifting working loads.

Crawler crane—A crane which consists of a rotating superstructure with power plant, operating machinery, and boom, mounted on a base, equipped with crawler treads for travel. Its function is to hoist and swing loads at various radii.

Designated—A person selected or assigned by the employer or the employer's representative as being qualified to perform specific duties.

Drum—The cylindrical members around which ropes are wound for raising and lowering the load or boom.

Dynamic (loading)—The loads introduced into the machine or its components by forces in motion.

Gantry (A-frame)—A structural frame, extending above the superstructure, to which the boom support ropes are reeved.

Jib— An extension attached to the boom point to provide added boom length for lifting specified loads. The jib may be in line with the boom or offset to various angles.

Load (working)—The external load, in pounds, applied to the crane, including the weight of load-attaching equipment such as load blocks, shackles, and slings.

Load block (upper)—The assembly of hook or shackle, swivel, sheaves, pins, and frame suspended from the boom point.

Load block (lower)—The assembly of hook or shackle, swivel, sheaves, pins, and frame suspended by the hoisting ropes.

Load hoist —A hoist drum and rope reeving system used for hoisting and lowering loads.

Load ratings—The crane's ratings in pounds established by the manufacturer in accordance with OSHA regulations.

Locomotive crane— A rotating superstructure with power plant, operating machinery and boom, mounted on a base or car equipped for travel on railroad track. It may be self-propelled or propelled by an outside source. Its function is to hoist and swing loads at various radii.

Outriggers —The extendable or fixed metal arms, attached to the mounting base, which rest on supports at the outer ends.

Rail clamp —A tong-like metal device, mounted on a locomotive crane car, which can be connected to the track.

Reeving—A rope system in which the rope travels around drums and sheaves.

Rope—Refers to a wire rope unless otherwise specified.

Side loading—A load applied at an angle to the vertical plane of the boom.

Sandby crane—A crane which is not in regular service but which is used occasionally or intermittently as required.

Standing (guy) rope—A supporting rope which maintains a constant distance between the points of attachment to the two components connected by the rope.

Structural competence—The ability of the machine and its components to withstand the stresses imposed by applied loads.

Superstructure—The rotating upper frame structure of the machine and the operating machinery mounted thereon.

Swing—The rotation of the superstructure for movement of loads in a horizontal direction about the axis of rotation.

Swing mechanism—The machinery involved in providing rotation of the superstructure.

Tackle—An assembly of ropes and sheaves arranged for hoisting and pulling.

Transit —The moving or transporting of a crane from one jobsite to another.

Travel —The function of the machine moving from one location to another on a jobsite.

Travel mechanism—The machinery involved in providing travel.

Truck crane—A rotating superstructure with powerplant, operating machinery and boom, mounted on an automotive truck equipped with a powerplant for travel. Its function is to hoist and swing loads at various radii.

Wheel mounted crane (wagon crane)—A rotating superstructure with powerplant, operating machinery and boom, mounted on a base or platform equipped with axles and rubber-tired wheels for travel. The base is usually propelled by the engine in the superstructure, but it may be equipped with a separate engine controlled from the superstructure. Its function is to hoist and swing loads at various radii.

Wheelbase—The distance between centers of front and rear axles. For a multiple axle assembly the axle center for wheelbase measurement is taken as the midpoint of the assembly.

Whipline (auxiliary hoist)—A separate hoist rope system of lighter load capacity and higher speed than provided by the main hoist.

Winch head—A power driven spool for handling of loads by means of friction between fiber or wire rope and spool.

DERRICKS (29 CFR 1910.181)

A-frame derrick—A derrick in which the boom is hinged from a cross member between the bottom ends of two upright members spread apart at the lower ends and joined at the top; the boom point secured to the junction of the side members, and the side members are braced or guyed from this junction point.

Basket derrick—A derrick without a boom, similar to a gin pole, with its base supported by ropes attached to corner posts or other parts of the structure. The base is at a lower elevation than its supports. The location of the base of a basket derrick can be changed by varying the length of the rope supports. The top of the pole is secured with multiple reeved guys to position the top of the pole to the desired location by varying the length of the upper guy lines. The load is raised and lowered by ropes through a sheave or block secured to the top of the pole.

Boom—A timber or metal section or strut, pivoted or hinged at the heel (lower end) at a location fixed in height on a frame or mast or vertical member, and with its point (upper end) supported by chains, ropes, or rods to the upper end of the frame, mast, or vertical member. A rope for raising and lowering the load is reeved through sheaves or a block at the boom point. The length of the boom shall be taken as the straight line distance between the axis of the foot pin and the axis of the boom point sheave pin, or where used, the axis of the upper load block attachment pin.

Boom harness—The block and sheave arrangement on the boom point to which the topping lift cable is reeved for lowering and raising the boom.

Boom point—The outward end of the top section of the boom.

Breast derrick—A derrick without boom. The mast consists of two side members spread farther apart at the base than at the top and tied together at top and bottom by rigid members. The mast is prevented from tipping forward by guys connected to its top. The load is raised and lowered by ropes through a sheave or block secured to the top crosspiece.

Chicago boom derrick—A boom which is attached to a structure, an outside upright member of the structure serving as the mast, and the boom being stepped in a fixed socket clamped to the upright. The derrick is complete with load, boom, and boom point swing line falls.

Derrick—An apparatus consisting of a mast or equivalent member held at the head by guys or braces, with or without a boom, for use with a hoisting mechanism and operating ropes.

Derrick bullwheel—a horizontal ring or wheel, fastened to the foot of a derrick, for the purpose of turning the derrick by means of ropes leading from this wheel to a powered drum.

Eye—A loop formed at the end of a rope by securing the dead end to the live end at the base of the loop.

Fiddle block—A block consisting of two sheaves in the same plane held in place by the same cheek plates.

Foot bearing or foot block (sill block)—The lower support on which the mast rotates.

Gin pole derrick—A derrick without a boom. Its guys are so arranged from its top as to permit leaning the mast in any direction. The load is raised and lowered by ropes reeved through sheaves or blocks at the top of the mast.

Gudgeon pin—A pin connecting the mast cap to the mast allowing rotation of the mast.

Guy—A rope used to steady or secure the mast or other member in the desired position.

Guy derrick—A fixed derrick consisting of a mast capable of being rotated, supported in a vertical position by guys, and a boom whose bottom end is hinged or pivoted to move in a vertical plane with a reeved rope between the head of the mast and the boom point for raising and lowering the boom, and a reeved rope from the boom point for raising and lowering the load.

Load, working—The external load, in pounds, applied to the derrick, including the weight of load attaching equipment such as load blocks, shackles, and slings.

Load block, lower—The assembly of sheaves, pins, and frame suspended by the hoisting rope.

Load block, upper—The assembly of sheaves, pins, and frame suspended from the boom.

Mast—The upright member of the derrick.

Mast cap (spider)—The fitting at the top of the mast to which the guys are connected.

Reeving—A rope system in which the rope travels around drums and sheaves.

Rope—Refers to wire rope unless otherwise specified.

Safety hook—A hook with a latch to prevent slings or load from accidentally slipping off the hook.

Side loading—A load applied at an angle to the vertical plane of the boom.

Sill—A member connecting the foot block and stiffleg or a member connecting the lower ends of a double member mast.

Shearleg derrick— A derrick without a boom and similar to a breast derrick. The mast, wide at the bottom and narrow at the top, is hinged at the bottom and has its top secured by a multiple reeved guy to permit handling loads at various radii by means of load tackle suspended from the mast top.

Standby derrick—A derrick not in regular service which is used occasionally or intermittently as required.

Stiffleg—A rigid member supporting the mast at the head.

Stiffleg derrick—A derrick similar to a guy derrick except that the mast is supported or held in place by two or more stiff members, called stifflegs, which are capable of resisting either tensile or compressive forces. Sills are generally provided to connect the lower ends of the stifflegs to the foot of the mast.

Swing—A rotation of the mast and/or boom for movements of loads in a horizontal direction about the axis of rotation.

FLAMMABLE AND COMBUSTIBLE LIQUIDS (29 CFR 1910.106):

Aerosol—A material which is dispensed from its container as a mist, spray, or foam by a propellant under pressure.

ASTM—The American Society for Testing and Materials

Atmospheric tank—A storage tank which has been designed to operate at pressures from atmospheric through 0.5 p.s.i.g.

Automotive service station—The portion of property where flammable or combustible liquids used as motor fuels are stored and dispensed from fixed equipment into the fuel tanks of motor vehicles and shall include any facilities available for the sale and service of tires, batteries, and accessories, and for minor automotive maintenance work. Major automotive repairs, painting, body and fender work are excluded.

Barrel—A volume of 42 U.S. gallons.

Boiling point—The boiling point of a liquid at a pressure of 14.7 pounds per square inch absolute (p.s.i.a.) (760 mm.). Where an accurate boiling point is unavailable for the material in question, or for mixtures which do not have a constant boiling point, the 10 percent point of a distillation performed in accordance with the Standard Method of Test for Distillation of Petroleum Products, ASTM D-86-62, which is incorporated by reference as specified in 1910.6, may be used as the boiling point of the liquid.

Boilover—The expulsion of crude oil (or certain other liquids) from a burning tank. The light fractions of the crude oil burnoff producing a heat wave in the residue, which on reaching a water strata may result in the expulsion of a portion of the contents of the tank in the form of froth.

Bulk plant —The portion of a property where flammable or combustible liquids are received by tank vessel, pipelines, tank car, or tank vehicle, and are stored or blended in bulk for the purpose of distributing such liquids by tank vessel, pipeline, tank car, tank vehicle, or container.

Chemical plant—A large integrated plant or that portion of such a plant other than a refinery or distillery where flammable or combustible liquids are produced by chemical reactions or used in chemical reactions.

Closed container—A container as herein defined, so sealed by means of a lid or other device that neither liquid nor vapor will escape from it at ordinary temperatures.

Combustible liquid—Any liquid having a flashpoint at or above 100°F (37.8°C) Combustible liquids shall be divided into two classes as follows:

(i) Class II liquids shall include those with flashpoints at or above 100°F (37.8°C) and below 140°F (60°C), except any mixture having components with flashpoints of 200°F (93.3°C) or higher, the volume of which makes up 99 percent or more of the total volume of the mixture.

(ii) Class III liquids shall include those with flashpoints at or above 140°F (60°C) Class III liquids are subdivided into two subclasses:

(a) Class IIIA liquids shall include those with flashpoints at or above 140°F (60°C) and below 200°F (93.3°C), except any mixture having components with flashpoints of 200°F (93.3°C), or higher, the total volume of which makes up 99 percent or more of the total volume of the mixture.

(b) Class IIIB liquids shall include those with flashpoints at or above 200°F (93.3°C). This section does not cover Class IIIB liquids. Where the term "Class III liquids" is used in this section, it shall mean only Class IIIA liquids.

(iii) When a combustible liquid is heated for use to within 30°F (16.7°C) of its flashpoint, it shall be handled in accordance with the requirements for the next lower class of liquids.

Container—Any can, barrel, or drum.

Crude petroleum—A hydrocarbon mixture that has a flash point below 150° F and which has not been processed in a refinery.

Distillery—A plant or that portion of a plant where flammable or combustible liquids produced by fermentation are concentrated, and where the concentrated products may also be mixed, stored, or packaged.

Fire area—An area of a building separated from the remainder of the building by construction having a fire resistance of at least 1 hour and having all communicating openings properly protected by an assembly having a fire resistance rating of at least 1 hour.

Flammable aerosol—An aerosol which is required to be labeled "Flammable" under the Federal Hazardous Substances Labeling Act (15 U.S.C. 1261). Such aerosols are considered Class IA liquids.

Flammable liquid—Any liquid having a flashpoint below 100°F (37.8°C), except any mixture having components with flashpoints of 100°F (37.8°C) or higher, the total of which makes up 99 percent or more of the total volume of the mixture. Flammable liquids shall be known as Class I liquids. Class I liquids are divided into three classes as follows:

(i)   Class IA shall include liquids having flashpoints below 73°F (22.8°C) and having a boiling point below 100°F (37.8°C).

(ii) Class IB shall include liquids having flashpoints below 73°F (22.8°C) and having a boiling point at or above 100°F (37.8°C).

(iii) Class IC shall include liquids having flashpoints at or above 73°F (22.8°C) and below 100°F (37.8°C).

Flashpoint—The minimum temperature at which a liquid gives off vapor within a test vessel in sufficient concentration to form an ignitable mixture with air near the surface of the liquid, which shall be determined as follows:

(i)   For a liquid which has a viscosity of less than 45 SUS at 100°F (37.8°C), does not contain suspended solids, and does not have a tendency to form a surface film while under test, the procedure specified in the Standard Method of Test for Flashpoint by Tag Closed Tester (ASTM D-56-70), which is incorporated by reference as specified in 1910.6, shall be used.

(ii) For a liquid which has a viscosity of 45 SUS or more at 100°F (37.8°C), or contains suspended solids, or has a tendency to form a surface film while under test, the Standard Method of Test for Flashpoint by Pensky-Martens Closed Tester (ASTM D-93-71) shall be used, except that the methods specified in Note 1 to 1.1 of ASTM D-93-71 may be used for the respective materials specified in the Note. The preceding ASTM standards are incorporated by reference as specified in 1910.6.

(iii) For a liquid that is a mixture of compounds that have different volatilities and flashpoints, its flashpoint shall be determined by using the procedure specified in paragraph (a)(14) (i) or (ii) of this section on the liquid in the form it is shipped. If the flashpoint, as determined by this test, is 100°F (37.8°C) or higher, an additional flashpoint determination shall be run on a sample of the liquid evaporated to 90 percent of its original volume, and the lower value of the two tests shall be considered the flashpoint of the material.

(iv) Organic peroxides, which undergo autoaccelerating thermal decomposition, are excluded from any of the flashpoint determination methods specified in this subparagraph.

Liquid —Any material which has a fluidity greater than that of 300 penetration asphalt when tested in accordance with ASTM Test for Penetration for Bituminous Materials, D-5-65, which is incorporated by reference as specified in 1910.6.When not otherwise identified, the term "liquid" shall include both flammable and combustible liquids.

Lower flammable limit—The percentage of vapor in the air below which a fire cannot occur because there is not enough fuel: the mixture is said to be too lean.

Low-pressure tank —A storage tank which has been designed to operate at pressures above 0.5 p.s.i.g. but not more than 15 p.s.i.g.

Marine service station—The portion of a property where flammable or combustible liquids used as fuels are stored and dispensed from fixed equipment on shore, piers, wharves, or floating docks into the fuel tanks of self-propelled craft, which shall include all facilities used in connection therewith.

Portable tank—A closed container having a liquid capacity over 60 U.S. gallons and not intended for fixed installation.

Pressure vessel—A storage tank or vessel which has been designed to operate at pressures above 15 p.s.i.g.

Protection from exposure—An adequate fire protection for structures on property adjacent to tanks, where there are employees of the establishment.

Refinery—A plant in which flammable or combustible liquids are produced on a commercial scale from crude petroleum, natural gasoline, or other hydrocarbon sources.

Safety can—An approved container, of not more than 5 gallons capacity, having a spring-closing lid and spout cover and so designed that it will safely relieve internal pressure when subjected to fire exposure.

Storage—Flammable or combustible liquids shall be stored in a tank or in a container that complies with the regulatory requirements.

SUS—Saybolt Universal Seconds, as determined by the Standard Method of Test for Saybolt Viscosity (ASTM D-88-56), which may be determined by use of the SUS conversion tables specified in ASTM Method D2161-66 following determination of viscosity in accordance with the procedures specified in the Standard Method of Test for Viscosity of Transparent and Opaque Liquids (ASTM D445-65).

Unstable (reactive) liquid —A liquid which in the pure state or as commercially produced or transported will vigorously polymerize, decompose, condense, or will become self-reactive under conditions of shocks, pressure, or temperature.

Upper flammable limit—The percentage of vapor in the air above which there is not enough air for a fire: the mixture is said to be too rich.

Vapor density—The weight of a flammable vapor compared to air (Air = 1). Vapors with a high density are more dangerous and require better ventilation because they tend to flow along the floor and collect in low spots.

Vapor pressure—The pressure, measured in pounds per square inch (absolute), exerted by a volatile liquid as determined by the "Standard Method of Test for Vapor Pressure of Petroleum Products (Reid Method)."

Ventilation—For the prevention of fire and explosion, it is considered adequate if it is sufficient to prevent accumulation of significant quantities of vapor-air mixtures in concentration over one-fourth of the lower flammable limit.

Viscous—A viscosity of 45 SUS or more.

HYDROGEN (29 CFR 1910.103)

Approved—Means, unless otherwise indicated, listed or approved by a nationally recognized testing laboratory. Refer to 1910.7 for definition of nationally recognized testing laboratory.

ASME—American Society of Mechanical Engineers.

DOT specifications—Regulations of the Department of Transportation published in 49 CFR Chapter I.

Gaseous hydrogen system—a system in which the hydrogen is delivered, stored, and discharged in the gaseous form to consumer's piping. The system includes stationary or movable containers, pressure regulators, safety relief devices, manifolds, interconnecting piping and controls. The system terminates at the point where hydrogen at service pressure first enters the consumer's distribution piping. [29 CFR 1910.103 (a)(1)]

POWERED INDUSTRIAL TRUCKS (29 CFR 1910.178)

Attachment—A device other than conventional forks or load backrest extension, mounted permanently or removable on the elevating mechanism of a truck for handling the load.

Popular types are fork extension clamps, rotating devices, side shifters, load stabilizers, rams, and booms.

Battery-electric truck—An electric truck in which the power source is a storage battery.

Cantilever truck—A self-loading counterbalanced or noncounterbalanced truck equipped with cantilever load engaging means, such as forks.

Carriage—A support structure for forks or attachments, generally rollermounted, traveling vertically within the mast of a cantilever truck.

Center Control—An operator-control position is located near the center of the truck.

Counterbalanced truck—A truck equipped with load-engaging means wherein all the load during normal transporting is external to the polygon formed by the wheel contacts.

Diesel-electric truck—An electric truck in which the power source is a diesel-engine-driven generator.

Dockboard—A portable or fixed device for spanning the gap or compensating for difference in level between loading platforms and carriers.

Electirc truck—A truck in which the principal energy is transmitted from power source to motor(s) in the form of electricity.

End control —The operator-control position is located at the end opposite the load end of the truck.

Forks —Horizontal tine—like projections, normally suspended from the carriage, for engaging and supporting loads.

Fork height—The vertical distance from the floor to the load-carrying surface adjacent to the heel of the forks with mast vertical, and in the case of reach trucks, with the forks extended.

Forklift truck—A high lift self-loading truck, equipped with load carriage and forks, for transporting and tiering loads.

Gas-electric truck—An electric truck in which the power source is a gasoline or LP gas-engine-driven generator.

High-lift truck—A self-loading truck equipped with an elevating mechanism designed to permit tiering.

High-lift platform truck—A self-loading truck equipped with an elevating mechanism intended primarily for transporting and tiering loaded skid platforms.

Industrial tractor—A powered industrial vehicle designed primarily to draw one or more nonpowered trucks, trailers, or other mobile loads.

Internal combustion engine truck—A truck in which the power source is a gas or diesel engine.

Load backrest extension—A device extending vertically from the fork carriage frame.

Load center (forklift)—the horizontal longitudinal distance from the intersection of the horizontal load-carrying surfaces and vertical load-engaging faces of the forks (or equivalent load-positioning structure) to the center of gravity of the load.

Narrow-aisle truck—A self-loading truck primarily intended for right-angle stacking in aisles narrower than those normally required by counterbalance trucks of the same capacity.

Order-picker truck high lift—A high-lift truck controllable by the operator stationed on a platform movable with the load-engaging means and intended for (manual) stock selection. The truck may be capable of selfloading and/or tiering.

Overhead guard—A framework fitted to a truck over the head of a riding operator.

Pallet truck—A self-loading, low-lift truck equipped with wheeled forks of dimensions to go between the top and bottom boards of a double-faced pallet, and having wheels capable of lowering into spaces between the bottom boards, so as to raise the pallet off the floor for transporting.

Parking brake—A device to prevent the movement of a stationary vehicle.

Powered industrial truck—A mobile, power-driven vehicle used to carry, push, pull, lift, or stack material.

Qualified operator—One who has had appropriate and approved training, including satisfactory completion of both written and operational tests to demonstrate knowledge and skill in the safe operation of the equipment to be used.

Reach truck—A self-loading truck, generally high-lift, having load engaging means mounted so it can be extended forward under control to permit a load to be picked up and deposited in the extended position and transported in the retracted position.

Rider truck—A truck that is designed to be controlled by a riding operator.

Side loader—A self-loading truck, generally high-lift, having load engaging means mounted in such a manner that it can be extended laterally under control to permit a load to be picked up and deposited in the extended position and transported in the retracted position.

Tiering—The process of placing one load on or above another.

RIGGING (29 CFR 1926.251)

Abrasion—Surface wear.

Alternate lay—Lay of wire rope in which the strands are alternately regular and Lang Lay.

Area, metallic — Sum of the cross sectional areas of individual wires in a wire rope or strand.

Basket or socket—The conical portion of a socket into which a splayed rope end is inserted and secured with zinc.

Becket loop—A loop of small rope or strand fastened to the end of a large wire rope to facilitate installation.

Bending stress—Stress imposed on wires of a wire rope by bending. This stress need not be added to direct load stresses. When sheaves and drums are of suitable size, it does not affect the normal life of the wire rope.

Breaking strength—The measured load required to break a cable or chain.

Bridle sling—A two-part sling attached to a single part line. Iee legs of the sling are spread to divide and equalize the load.

Bull ring—The main large ring of a sling to which sling legs are attached.

Cable—A term loosely applied to wire ropes, wire strands, manila ropes, and electrical conductor.

Cable Laid Wire Rope—A type of wire rope consisting of several independent wire ropes laid into a single wire rope.

Cable crowd rope—A wire rope used to force the bucket of a power shovel into the material being handled.

Center—A single wire or fiber in the center of a strand about which the wires are laid.

Choker rope—A short wire rope sling used to form a slip noose around the object to be moved or lifted.

Circumference—Measured perimeter of a circle circumscribing the wires of a strand or the strands of a wire rope.

Clamp, strand —A fitting for forming a loop at the end of a length of strand consisting of two grooved plates and bolts.

Clevis—A U-shaped fitting with pins.

Clip—A fitting for clamping two parts of wire rope.

Closed socket—A wire rope fitting consisting of an integral becket and bail.

Coil—A circular bundle of wire rope not packed on a reel.

Come along—A device for making a temporary grip and exerting tension on a wire rope.

Continuous bend—Reeving of wire rope over sheaves and drums so that it bends in one direction, as opposed to reverse bend.

Core—Core member of a wire about which the strands are laid. It may be fiber, a wire strand, or an independent wire rope.

Corrosion—Chemical decomposition by exposure to moisture, acids, alkalis, or other destructive agents.

Critical, diameter—A diameter of the smallest bend for a given wire rope which permits the wires and strands to adjust themselves by relative movement while remaining in their normal position.

Critical service—The use of equipment or tackle for hoisting, rigging, or handling of High-Consequence or Special High-Consequence Loads.

Deceleration stress—Additional stress imposed on a wire rope due to decreasing the velocity of the load.

Deflection—Sag of a rope in a span. Usually measured at midspan as the depth from a chord joining the tops of the two supports and any deviation from a straight line,

Diameter—Distance measured across the center of a circle circumscribing the wires of a strand or the strands of a wire rope.

Dog-leg—Permanent short bend or kink in a wire rope caused by improper use.

Elastic limit—Limit of stress above which a permanent deformation takes place within the material. This limit is approximately 55-65% of breaking strength of steel-wire ropes.

Equalizing slings—Slings composed of wire rope and equalizing fittings.

Equalizing thimbles—Special type of fitting used as a component part of some wire rope slings.

Eye or eye splice—A loop with or without a thimble formed in the end of a wire rope.

Factor of safety—Ratio of ultimate strength to the design working stress.

Fiber centers—Cords or rope made of vegetable fiber used in the center of a strand.

Fiber cores—Cords or rope made of vegetable fiber used in core of a wire rope.

Fitting—Any accessory used as an attachment for wire rope.

Flat rope—Wire rope made of parallel alternating right lay and left lay ropes sewn together by relatively soft wires.

Flattened strand rope—A wire rope with either oval or triangular shaped strands which present a flattened rope surface.

Galvanized—To coat with zinc to protect against corrosion.

Galvanized rope—Rope made of galvanized wire.

Galvanized strand—Strand made of galvanized wire.

Galvanized wire—Wire coated with zinc.

Grommet—An endless 7-strand wire rope made from one continuous length of strand.

Hook load—The total live load supported by the hook of a crane, derrick, or other hoisting equipment, including the load, slings, spreader bars, and other tackle not part of the load but supported by the hook and required for handling of the load.

Independent wire rope core—Wire rope used as the core of a larger rope.

Inner wires—All wires of a strand except surface or cover wires.

Internally lubricated—Wire rope or strand having all wires coated with lubricant.

Kink—Permanent distortion of the wires and strands resulting from sharp bends.

Lagging—External wood covering on a reel of rope or strand.

Lang lay rope—Wire rope in which the wires in the strands and the strands in the rope are laid in the same directions.

Lay—The lengthwise distance on a wire rope in which a strand makes one complete turn around the rope.

Left lay:
(a) Strand: Stand in which the cover wires are laid in a helix having a left hand pitch, similar to a left hand screw.
(b) Rope: Rope in which the strands are laid in a helix having a left hand pitch, similar to a left hand screw.

Right Lay:

(a)   Strand: Strand in which the cover wires are laid in a helix having a right hand pitch, similar to a right hand screw,

(b)   Rope: Rope in which the strands are laid in a helix having a right hand pitch, similar to a right hand screw.

Load limit—Refers to the maximum allowed weight.

Marline spike—Tapered steel pin used in splicing wire rope.

Mousing—A method of bridging the throat opening of a hook to prevent the release of load lines and slings, under service or slack conditions, by wrapping with soft wire, rope, heavy tape, or similar materials.

Nonrotating wire rope—8 x 7 wire rope consisting of a left lay, lang lay inner rope covered by 12 seven wire strands right lay, regular lay.

Open socket—Wire rope fitting consisting of a "basket" and two "ears" with a pin.

Peening—Permanent distortion of outside wire in a rope caused by pounding.

Preformed wire rope—Wire rope in which the strands are permanently shaped, before fabrication into the rope, to the helical form they assume in the wire rope.

Preformed strand—Strand in which the wires are permanently shaped, before fabrication in the strands, to the helical form they assume in the strand.

Prestressing—Stressing a wire rope or strand before use under such a tension and for such a time that the constructional stretch is largely removed.

Qualified rigger—One whose competence in this skill has been demonstrated by experience satisfactory to the appointed person. A worker extensively experienced including rigging and handling of items.

Regular lay rope—Wire rope in which the wires in the strands and the strands in the rope are laid in opposite directions.

Rigging equipment—The devices used to lift material, equipment or other objects such as chain slings, cable slings, rope or web slings and special fixtures.

Seize—To bind securely the end of a wire rope or strand with seizing wire or strand.

Seizing strand—Small strand, usually of seven wires, made of soft annealed iron wire.

Serve—To cover the surface of a wire rope or strand with a wrapping of wire.

Shackle—A type of clevis normally used for lifting.

Slings—Wire ropes or chains made into forms, with or without fittings, for handling loads.

Slings, Braided—A very flexible sling composed of several individual wire ropes braided into a single sling.

Splicing—Interweaving of two ends of ropes so as to make a continuous or endless length without appreciably increasing the diameter. Also making a loop or eye in the end of a rope by tucking the ends of the strands.

Stainless steel rope—Wire rope made of chrome-nickel steel wires having great resistance to corrosion.

Steel clad rope—Rope with individual strands spirally wrapped with flat steel wire.

Stirrup—The U-bolt or eyebolt attachment on a bridge socket.

Strand—An arrangement of wires helically laid about an axis, or another wire or fiber center to produce a symmetrical section.

Swaged fitting—Fittings in which wire rope is inserted and attached by a cold forming method.

Tappering and welding—Reducing the diameter of the end of a wire rope and welding it to facilitate reeving.

Thimble—Grooved-metal fitting to protect the eye of a wire rope.

Turnbuckle—Device attached to wire rope for making limited adjustments in length. It consists of a barrel and right and left hand threaded bolts.

Wedge socket—Wire rope fitting in which the rope end is secured by a wedge.

Wire rope—A plurality of strands laid helically around an axis or a core.

ROBOTS:

Actuator—A power mechanism used to effect motion of the robot or a device that converts electrical, pneumatic, or hydraulic energy into robot motion.

Application program—The instruction set that defines the specific tasks of robots and robot systems. This program may be originated and/or modified by the robot operator.

Attended continuous operation—The time when robots are performing production tasks at a speed no greater than slow speed through attended program execution.

Attended program verification—The time when a person within the restricted envelope verifies the robot's programmed tasks at programmed speed.

Automatic mode—The state of the robot in which automatic operation can be initiated.

Automatic operation—The time during which robots are performing programmed tasks through unattended program execution.

Awareness barrier—Physical and/or visual means that warn a person of an approaching or present hazard.

Awareness signal—A device that warns a person of an approaching or present hazard by means of audible sound or visible light.

Backfit—A change either to the hardware or design applied to systems, structures, and components; to a procedure or organization used to design, construct, or operate a facility; or to any contractually agreed to activity that may be required because of new or other safety requirements.

Barrier—A physical means of separating persons from the restricted envelope.

Control device—A control hardware providing a means for human intervention in the control of a robot or robot system, such as an emergency stop button, start button, or selector switch.

Control program—An inherent set of control instructions that defines the actions and responses of the robot system. This program is usually not intended to be modified by the user.

Controller—An information-processing device whose inputs are both desired and measured positions, velocity or other variables in a process, and whose output is a drive signal to a controlling motor or actuator.

Degrees of freedom—The directions of motion inherent in the design of the robot mechanical system. Robot systems can have up to six degrees of freedom or types of movement These are:

- Vertical traverse — up and down motion of the robot arm caused by pivoting the entire arm about a horizontal axis or moving the arm along a vertical slide.
- Radial traverse—retraction and extension of the arm (in and out movements).
- Rotational traverse—rotation about the vertical axis (right or left swivel of the robot arm).
- Wrist pitch—up and down movement of the wrist.
- Wrist yaw—right and left swivel of the wrist.
- Wrist roll—rotation of the wrist.

Emergency stop—The operation of a circuit with hardware-based components that overrides all other robot controls, removes drive power from the robot actuators, and causes all moving parts to stop.

Enabling device—A manually operated device that, when continuously activated, permits motion. Releasing pressure on this device stops robot motion and motion of associated equipment that may present a hazard.

End effector—An accessory tool or device specifically designed for attachment to the robot wrist or tool-mounting plate to enable the robot to perform its intended task. (Examples may include gripper, spot-weld gun, are-weld gun, spray paint gun, or any other application tools.)

Hazard—A situation that could cause physical harm or have adverse safety consequences.

Hazardous motion—Any motion that could cause physical harm to personnel or equipment.

Human engineering—The application of available knowledge that defines the nature and limits of human capabilities as they relate to checkout, operation, maintenance, or control of systems or equipment in engineering design.

Industrial equipment—A physical apparatus used to perform industrial tasks (e.g., welders, conveyors, turntables, positioning tables, machine tools, fork trucks, or robots).

Industrial robot—A reprogrammable, multifunctional manipulator designed to move material, parts, tools, or specialized devices through variable programmed motions. This includes industrial robot systems, end effectors, and devices and sensors.

Industrial robot system—The system includes not only industrial robots but also any devices and/or sensors required for the robot to perform its tasks, including communication interfaces for sequencing or monitoring the robot. See Appendix B for more detailed definition.

Interlock—A situation in which the operation of one control or mechanism causes or prevents the operation of another.

Limiting device—A device that restricts the maximum envelope by causing all robot motion to stop and is independent of the control and application programs.

Maximum envelope—The volume of space encompassing the maximum-designed movements of all robot parts. This includes the workpiece, end effector, and attachments.

Mobile robot—A self-propelled and self-contained robot that can move over a mechanically unconstrained course.

Muting—The deactivation of a presence-sensing safeguarding device during part of the robot cycle.

Operating envelope—That part of the restricted envelope (space) used by the robot while performing its programmed motions.

Operator—A person designated to start, monitor, and stop the intended operation of a robot or robot system.

Pendant—Any portable control device, including teach pendants, that permits an operator to control the robot from within or without the restricted envelope of the robot.

Presence-sensing safeguarding device—A device designed, constructed, and installed to create a sensing field or area to detect an intrusion into such field or area by personnel, robots, or other objects.

Rebuild—To restore, to the extent possible, the robot to the original manufacturers' specifications.

Repair—To restore robots and robot systems to operating condition after malfunction, damage, or wear.

Restricted envelope—That part of the maximum envelope to which a robot is restricted by limiting devices. The boundaries of the restricted envelope are defined by the maximum distance that the robot and associated tooling can travel after the limiting device is actuated. Note: The safeguarding interlocking logic and the robot program may allow the restricted envelope to be continually redefined (dynamic restricted envelope) while the robot performs its application program.

Robot manufacturer—An entity involved in either the design, building, or sale of robots, robot tooling, robotic peripheral equipment or controls, and/or associated process equipment.

Robot system integrator—An entity that either directly or through subcontractors assumes responsibility for the design, building, and/or integration of the robot or ancillary equipment for a robotic application.

Safeguard—A barrier guard, device, or safety procedure designed for the protection of personnel. "Safeguard" is not used in its nuclear material security sense in this chapter.

Safety procedure—Documented instructions designed for the protection of personnel and equipment.

Selective compliance assembly robot arm (SCARA)—A robot manipulator geometry, typically with four degrees of freedom that includes two deep arms connected by vertical pin hinges to provide X-Y (horizontal plane) motion with high rigidity. The first and second links of the base allow for rotation along the horizontal plane, while vertical plane motion is typically provided by a mast which the robot travels up and down on.

Service or personal robots—Robots that are typically used for educational purposes or for the training of industrial robot operators and maintenance personnel.

Single point of control—The ability to operate the robot such that initiation or robot motion from one source of control is only possible from that source and cannot be overridden by another source.

Slow speed control—A mode of robot motion control in which the velocity of the robot is limited to 6 inches (150 millimeters) per second to allow personnel sufficient time to either withdraw from the hazardous motion or stop the robot.

Startup—Routine application of drive power to the robot or robot system.

Startup, initial— Initial drive-power application to the robot or robot system after one of the following events:
- Manufacture or modification.
- Installation or reinstallation.
- Programming or program editing.
- Maintenance or repair.

Teach—The generation and storage of a series of positional data points affected by moving the robot arm through a path of intended motions.

Teach mode—The control state that allows for the generation and storage of positional data points affected by moving the robot arm through a path intended for the robot.

Teacher—A person who provides the robot with a specific set of instructions.

User — An entity, experimenter, or person that uses robots or contracts, hires, or is responsible for the personnel associated with the robot operation.

OVERHEAD AND GANTRY CRANES (29CFR 1910.179):

Appointed—A person assigned specific responsibilities by the employer or the employer's representative.

ANSI—The American National Standards Institute.

Automatic crane—A crane which when activated operates through a preset cycle or cycles.

Auxiliary hoist—A supplemental hoisting unit of lighter capacity and usually higher speed than provided for the main hoist.

Brake—A device used for retarding or stopping motion by friction or power means.

Bridge—That part of a crane consisting of girders, trucks, end ties, footwalks, and drive mechanism which carries the trolley or trolleys.

Bridge travel —The crane movement in a direction parallel to the crane runway.

Bumper (buffer)—An energy absorbing device for reducing impact when a moving crane or trolley reaches the end of its permitted travel; or when two moving cranes or trolleys come in contact.

Cab —the operator's compartment on a crane.

Cab-operated crane—A crane controlled by an operator in a cab located on the bridge or trolley.

Cantilever gantry crane—A gantry or semigantry crane in which the bridge girders or trusses extend transversely beyond the crane runway on one or both sides.

Clearance—The distance from any part of the crane to a point of the nearest obstruction.

Collectors, current—The contacting devices for collecting current from runway or bridge conductors.

Conductors, bridge—The electrical conductors located along the bridge structure of a crane to provide power to the trolley.

Conductors, runway (main)—The electrical conductors located along a crane runway to provide power to the crane.

Control braking—A method of controlling crane motor speed when in an overhauling condition.

Controller, spring return—A controller which when released will return automatically to a neutral position.

Countertorque—A method of control by which the power to the motor is reversed to develop torque in the opposite direction.

Crane—A machine for lifting and lowering a load and moving it horizontally, with the hoisting mechanism an integral part of the machine. Cranes, whether fixed or mobile, are driven manually or by power.

Cranes, types of:

    Automatic Crane—A crane which, when activated, operates through a preset cycle or cycles.

    Cab-Operated Crane—A crane controlled by an operator in a cab located on the bridge or trolley.

    Cantilever Gantry Crane—A gantry or semi-gantry crane in which the bridge girders or trusses extend transversely beyond the crane runway on one or both sides.

    Floor-Operated Crane — A crane which is pendant or nonconductive rope controlled by an operator on the floor or an independent platform.

    Gantry Crane—A crane similar to an overhead crane, except that the bridge for carrying the trolley or trolleys is rigidly supported on two or more legs running on fixed rails or other runway.

    Overhead Crane—A crane with a movable bridge carrying a movable- or fixed-hoisting mechanism and traveling on an overhead fixed runway structure.

    Power-Operated Crane—A crane whose mechanism is driven by electricity, air, hydraulic, or internal combustion.

    Pulpit-Operated Crane—A crane operated from a Reed operator station not attached to the crane.

    Remote-Operated Crane—A crane controlled by an operator not in a pulpit nor in a cab attached to the crane, by any method other than pendant or rope control (e.g., radio-controlled crane).

    Semi-Gantry Crane—A gantry crane with one end of the bridge rigidly supported on one or more legs that run on a fixed rail or runway, the other end of the bridge being supported by a truck running on an elevated rail or runway.

    Wall or Jib Crane—A crane having a jib with or without trolley, and supported from a side wall or line of columns of a building.

Drag brake—A brake which provides retarding force without external control.

Designated—A person selected or assigned by the employer or the employer's representative as being qualified to perform specific duties.

Drift point—A point on a travel motion controller which releases the brake while the motor is not energized. This allows for coasting before the brake is set.

Drum—The cylindrical member around which the ropes are wound for raising or lowering the load.

Dynamic—A method of controlling crane motor speeds when in the overhauling condition to provide a retarding force.

Emergency stop switch—A manually or automatically operated electric switch to cut off electric power independently of the regular operating controls.

Equalizer—A device which compensates for unequal length or stretch of a rope.

Exposed—Capable of being contacted inadvertently. Applied to hazardous objects not adequately guarded or isolated.

Fail-safe—A provision designed to automatically stop or safely control any motion in which a malfunction occurs.

Floor-operated crane—A crane which is pendant or nonconductive rope controlled by an operator on the floor or an independent platform.

Footwalk—The walkway with handrail, attached to the bridge or trolley for access purposes.

Gantry crane—A crane similar to an overhead crane, except that the bridge for carrying the trolley or trolleys is rigidly supported on two or more legs running on fixed rails or other runway.

Hoist—An apparatus which may be a part of a crane, exerting a force for lifting or lowering.

Hoist chain—The load bearing chain in a hoist. Note: Chain properties do not conform to those shown in ANSI B30.9-1971, Safety Code for Slings.

Hoist motion—That motion of a crane which raises and lowers a load.

Holding brake—A brake that automatically prevents motion when power is off.

Hot metal handling crane—An overhead crane used for transporting or pouring molten material.

Limit switch—A switch which is operated by some part or motion of a power-driven machine or equipment to alter the electric circuit associated with the machine or equipment.

Magnet—An electromagnetic device carried on a crane hook to pick up loads magnetically.

Main hoist—The hoist mechanism provided for lifting the maximum rated load.

Main switch—A switch controlling the entire power supply to the crane.

Man trolley—A trolley having an operator's cab attached thereto.

Master switch—A switch which dominates the operation of contactors, relays, or other remotely operated devices.

Mechanical—A method of control by friction.

Overhead crane—A crane with a movable bridge carrying a movable or fixed hoisting mechanism and traveling on an overhead fixed runway structure.

Power-operated crane—A crane whose mechanism is driven by electric, air, hydraulic, or internal combustion means.

Pulpit-operated crane—A crane operated from a fixed operator station not attached to the crane.

Rated load—The maximum load for which a crane or individual hoist is designed and built by the manufacturer and shown on the equipment nameplate(s).

Regenerative—A form of dynamic braking in which the electrical energy generated is fed back into the power system.

Remote-operated crane—A crane controlled by an operator not in a pulpit or in the cab attached to the crane, by any method other than pendant or rope control.

Rope—This refers to wire rope, unless otherwise specified.

Running sheave—A sheave which rotates as the load block is raised or lowered.

Runway—An assembly of rails, beams, girders, brackets, and framework on which the crane or trolley travels.

Semigantry crane—A gantry crane with one end of the bridge rigidly supported on one or more legs that run on a fixed rail or runway, the other end of the bridge being supported by a truck running on an elevated rail or runway.

Side pull—That portion of the hoist pull acting horizontally when the hoist lines are not operated vertically.

Span—The horizontal distance center to center of runway rails.

Special rated capacity—The maximum hook load which a piece of hoisting equipment, or the maximum working load which an industrial truck or piece or rigging tackle, is permitted to carry, based on its present condition and the operational conditions as determined by an engineering evaluation, load test, or both. The special rated capacity may be equal to but not greater than the rated capacity of equipment established by the manufacturer.

Standby crane—A crane which is not in regular service but which is used occasionally or intermittently as required.

Stop —A device to limit travel of a trolley or crane bridge. This device normally is attached to a fixed structure and normally does not have energy absorbing ability.

Storage bridge crane—A gantry type crane of long span usually used for bulk storage of material; the bridge girders or trusses are rigidly or nonrigidly supported on one or more legs. It may have one or more fixed or hinged cantilever ends.

Switch—A device for making, breaking, or for changing the connections in an electric circuit.

Trolley—The unit which travels on the bridge rails and carries the hoisting mechanism.

Trolley travel—The trolley movement at right angles to the crane runway.

Truck—The unit consisting of a frame, wheels, bearings, and axles which supports the bridge girders or trolleys.

Wall crane—A crane having a jib with or without trolley and supported from a side wall or line of columns of a building. It is a traveling type and operates on a runway attached to the side wall or columns.

SLINGS (29 CFR 1910.184)

Angle of loading—The inclination of a leg or branch of a sling measured from the horizontal or vertical plane as shown in Fig. N-184-5; provided that an angle of loading of five degrees or less from the vertical may be considered a vertical angle of loading.

Basket hitch—A sling configuration whereby the sling is passed under the load and has both ends, end attachments, eyes or handles on the hook or a single master link.

Braided wire rope—A wire rope formed by plaiting component wire ropes.

Bridle wire rope sling—A sling composed of multiple wire rope legs with the top ends gathered in a fitting that goes over the lifting hook.

Cable laid endless sling-mechanical joint—A wire rope sling made endless by joining the ends of a single length of cable laid rope with one or more metallic fittings.

Cable laid grommet-hand tucked—An endless wire rope sling made from one length of rope wrapped six times around a core formed by hand tucking the ends of the rope inside the six wraps.

Cable laid rope—A wire rope composed of six wire ropes wrapped around a fiber or wire rope core.

Cable laid rope sling-mechanical joint—A wire rope sling made from a cable laid rope with eyes fabricated by pressing or swaging one or more metal sleeves over the rope junction.

Choker hitch—A sling configuration with one end of the sling passing under the load and through an end attachment, handle, or eye on the other end of the sling.

Coating—An elastomer or other suitable material applied to a sling or to a sling component to impart desirable properties.

Cross rod —A wire used to join spirals of metal mesh to form a complete fabric. (See Fig. N-184-2.)

Designated—A person selected or assigned by the employer or the employer's representative as being qualified to perform specific duties.

Equivalent entity—A person or organization (including an employer) which, by possession of equipment, technical knowledge and skills, can perform with equal competence the same repairs and tests as the person or organization with which it is equated.

Fabric (metal mesh)—The flexible portion of a metal mesh sling, consisting of a series of transverse coils and cross rods.

Female handle (choker)—A handle with a handle eye and a slot of such dimension as to permit passage of a male handle, thereby allowing the use of a metal mesh sling in a choker hitch. (See Fig. N-184-1.)

Handle—A terminal fitting to which metal mesh fabric is attached. (See Fig. N-184-1.)

Handle eye—An opening in a handle of a metal mesh sling shaped to accept a hook, shackle or other lifting device. (See Fig. N-184-1.)

Hitch—A sling configuration whereby the sling is fastened to an object or load, either directly to it or around it.

Link—A single ring of a chain.

Male handle (triangle)—A handle with a handle eye.

Master coupling link—An alloy steel welded coupling link used as an intermediate link to join alloy steel chain to master links. (See Fig. N-184-3.)

Master link or gathering ring—A forged or welded steel link used to support all members (legs) of an alloy steel chain sling or wire rope sling. (See Fig. N-184-3.)

Mechanical coupling link—A nonwelded, mechanically closed steel link used to attach master links, hooks, etc., to alloy steel chain.

Proof load—The load applied in performance of a proof test.

Proof test—A nondestructive tension test performed by the sling manufacturer or an equivalent entity to verify construction and workmanship of a sling.

Rated capacity or working load limit—The maximum working load permitted by the provisions of this section.

Reach—The effective length of an alloy steel chain sling, measured from the top bearing surface of the upper terminal component to the bottom bearing surface of the lower terminal component.

Selvage edge—The finished edge of synthetic webbing designed to prevent unraveling.

Sling—An assembly which connects the load to the material handling equipment.

Sling manufacturer—A person or organization that assembles sling components into their final form for sale to users.

Spiral—A single transverse coil that is the basic element from which metal mesh is fabricated. (See Fig. N-184-2.)

Strand laid endless sling-mechanical joint—A wire rope sling made endless from one length of rope with the ends joined by one or more metallic fittings.

Strand laid grommet-hand tucked—An endless wire rope sling made from one length of strand wrapped six times around a core formed by hand tucking the ends of the strand inside the six wraps.

Strand laid rope—A wire rope made with strands (usually six or eight) wrapped around a fiber core, wire strand core, or independent wire rope core (IWRC).

Vertical hitch—Method of supporting a load by a single, vertical part or leg of the sling. (See Fig. N-184-4.)

INDEX

I

U

U-bolts, 140
Underhung,
    cranes, 68

V

Vehicles,
    forklifts, 179
    highway, 151
    powered industrials trucks, 179
    off-road, 165
Visibility, 102

W

Walking,
    surfaces, 27
Web slings, synthetic
    environmental conditions, 145
    inspection, 143
    possible defects, 145
    removal from service, 147
    repairs, 147
    safe operating temperatures, 147
    sling identification, 145
    web sling properties, 144
Welding and cutting,
    use of fuel gases, 257
Wire mesh,
    slings, 140
Wire rope, slings
    field lubrication, 139
    general information, 139
    inspections and maintenance, 136
    life, 138
    removal from service, 138
    rope lay, 134
    safe operating temperatures, 136
    selection
        abrasive wear, 139
        abuse, 130
        fatigue, 138
        strength, 138
    storage, 139
    U-bolt, 140
Work practices,
    forklifts, 184
Working,
    surfaces, 27
Worksheet,
    job analysis, 23